T0177116

# Darwin and
# His Bears

# Darwin and His Bears

## How Darwin Bear and
## His Galápagos Islands Friends
## Inspired a Scientific Revolution

## Frank J. Sulloway

ILLUSTRATED BY THE AUTHOR

Blast Books
NEW YORK

Published by Blast Books, Inc.
P. O. Box 51, Cooper Station
New York, NY 10276-0051
www.blastbooks.com

Book design by Fearn Cutler de Vicq and Laura Lindgren
Typeset in ITC New Baskerville and Saygon
Cover design by Laura Lindgren

ISBN: 978-0922233-51-9 (alk. paper)

Library of Congress Control Number: 2021936746

Printed and bound by INK69, Belgium

First Edition 2021

10 9 8 7 6 5 4 3 2 1

## To Edward O. Wilson

Whose deep love for the natural world
and far-reaching understanding
of its fundamental principles have enlightened us all.

# Contents

# Foreword

In June of 2004, I was honored to join Frank Sulloway on a month-long expedition to retrace Charles Darwin's footsteps in the Galápagos Islands. Darwin had visited the island archipelago in 1835, and owing to the fame that Darwin's theory of evolution brought to the Galápagos, much had happened there subsequently to change the delicate ecological balance. Sulloway wanted to document those changes, particularly through photographs from the past century, whose locations we worked to pinpoint for contemporary comparison to see the transformations caused by natural and human activity.

The trip was so physically grueling, it renewed my respect for what the young British naturalist accomplished there so long ago. Charles Darwin was not only one sagacious scientist; he was one tenacious explorer in the harsh, arid, bleak lava landscape of these precarious islands. After a brutal climb through a moonscape-like area Darwin called the "craterized district," Frank and I collapsed in exhaustion, drenched in sweat—and later in Darwin's diary we read his description of a similar excursion as "a long walk."

A historian of science and Darwin scholar, Frank Sulloway has spent decades reconstructing how Darwin pieced together the theory of evolution. The iconic myth endures that Darwin became an evolutionist while in the Galápagos, having observed the beaks of finches and the carapaces of tortoises, and how each species was uniquely adapted by diet or island ecology. But Sulloway has been able to date a striking passage about the Galápagos mockingbirds in Darwin's unpublished manuscripts to nine months after departing the Galápagos, and to give it a proper interpretation. "When I see these Islands in sight of each other, & possessed of but a scanty stock of animals, tenanted by these birds, but slightly differing in structure & filling the same place in Nature, I must suspect they are only varieties." Darwin was talking about similar *varieties* of fixed kinds, not the *evolution* of separate species. At the time, Darwin was still a believer in creationism—which explains why he did not bother to record the locations of the few finches he collected, and being unable to argue that geographic isolation had facilitated their evolution, why he never specifically mentioned these now-famous birds in the *Origin of Species*.

Through careful analysis, Sulloway dates Darwin's acceptance of evolution to the second week of March 1837, after a meeting Darwin had with the eminent ornithologist John Gould, who had been studying his Galápagos bird specimens, and who corrected a number of taxonomic errors Darwin had made.

Darwin left the meeting with Gould, Sulloway concludes, convinced "beyond a doubt that transmutation must be responsible for the presence of similar but distinct species on the different islands of the Galápagos group. The supposedly immutable 'species barrier' had finally been broken, at least in Darwin's own mind." It was not until a decade after he visited the Galápagos that Darwin wrote: "Hence both in space and time, we seem to be brought somewhat near to that great fact—that mystery of mysteries—the first appearance of new beings on this earth."

As Frank and I hiked and camped, I learned the story of the vital help Darwin had received from indigenous Galápagos bears that I was surprised never to have previously encountered in my own studies. In time, the story Frank told me evolved into *Darwin and His Bears*, the masterpiece you hold in your hands. Uniquely imaginative, erudite, and thoroughly entertaining, *Darwin and His Bears* captures the joy of science as it reveals the true story of Darwin's gradual comprehension of evolution by natural selection—one of the greatest ideas in the history of science.

Michael Shermer

# PART ONE

# An Extraordinary Story

Charles Darwin in 1840 (age thirty-one), four years after his circumnavigation of the globe on H.M.S. *Beagle* (1831–1836). (After a portrait by George Richmond)

## CHAPTER 1

# An Odd Encounter in the Library

—— or ——

# I Meet Darwin Bear
# and Hear an Extraordinary Story

This is the story of how I met Darwin Bear in the most unexpected of places. I had gone to a rare-books library at Harvard University to consult a copy of Charles Darwin's revolutionary book *On the Origin of Species by Means of Natural Selection*, which was published more than a hundred and fifty years ago, in 1859. To my disappointment, the book was already signed out, so I asked the librarian when it might be returned. "Oh, that gentleman over there wearing the academic cap is reading it." She pointed toward a table in the corner.

There sat a small, fluffy, tan-colored bear reading a book with the characteristic green binding of a first edition of Darwin's *Origin of Species*. "Do you mean that bear over there?" I asked the librarian.

"Well, I hadn't actually noticed that he was a bear. He seemed so scholarly."

Intrigued by the oddity of the scene, I couldn't help going over to the little bear to find out just what he was up to. The bear was completely absorbed in the book and didn't seem to notice me peering quizzically over his shoulder. Finally, I cleared my throat: "Ahem, . . . Why are you, a bear, reading Charles Darwin's *Origin of Species*?"

The little bear stared up at me with a look of considerable surprise: "Why, sir, don't you know that this book is all about bears?"

"All about bears!" I exclaimed incredulously—to which I added, as politely as I could: "I've personally read all of Darwin's books, and it's hardly my impression that this particular one is about bears."

Unperturbed, the perky little bear quickly thumbed through the library's first edition until he'd found what he was looking for—on page 184. "This very passage," he explained, "is the key to Darwin's whole book and to all of his ideas about evolution. Let me read it to you." Then, with a self-satisfied smile on his face, and with evident pride, the bear proceeded to read the following sentence aloud: "In North America the black bear was seen . . . swimming for hours with widely open mouth, thus catching, like a whale, insects in the water."

Here the little bear opened his mouth very wide, in imitation of the whale-bear in Darwin's text. He then resumed his reading:

Darwin Bear, imitating Darwin's text about insect-catching bears.

Even in so extreme a case as this, if the supply of insects were constant, and if better adapted competitors did not already exist in the country, I can see no difficulty in a race of bears being rendered, by natural selection, more and more aquatic in their structure and habits, with larger and larger mouths, till a creature was produced as monstrous as a whale.

After the bear had finished reciting this passage, he paused for a few seconds in silent satisfaction. "So you see," he added, "Darwin's *Origin of Species* is all about bears, just as I said. Most people don't appreciate this fact, because in the later editions of his book

Darwin Bear and me on our way to a café.

Darwin deleted his much-maligned assertion that a bear could actually evolve into a whale.*

"In any event," the little bear continued, "Darwin is telling us that given enough time, bears could evolve into whales, and hence that whales and other sea creatures—such as porpoises and dolphins—may once have been bearlike creatures that looked very much like me. After all, bears are known to be very clever, so it makes perfect sense that some of them should have evolved into creatures as different and majestic as whales."

---

* For further information about the fate of this controversial passage in Darwin's *Origin of Species*, see Appendix 3: The Untold Story of Darwin's Whale-Bear (pp. 137–43).

I could hardly argue with my little acquaintance's recitation of the well-known passage from the first edition of the *Origin of Species*. Nor could I fault his logic. I decided to ask him if he'd like to join me for some tea and biscuits in order to continue our conversation.

The little bear graciously accepted my invitation, so we returned the book and headed to a nearby café—the bear bringing with him a large white bag in which I assumed were his belongings.

"By the way," I asked him as we walked along, "what's your name?"

"You may call me Darwin Bear," he said, skipping merrily alongside me.

"Please call me Frank," I responded.

And so began a most unexpected friendship—and adventure.

Darwin Bear, seated on his white travel bag, having tea and biscuits with me at the café.

## CHAPTER 2

# Darwin Bear Reveals a Scientific Secret

—— or ——

# The Little Bear Tells Another Incredible Tale

"So, how is it that you know so much about Darwin's *Origin of Species*?" I asked as we sat down together to have some tea.

Perched on his large white bag—with a teacup balanced carefully on his lap—the little bear stared me right in the eye. "You may not believe this, but I knew Charles Darwin. In fact, I knew him very well. You might even say we were the best of friends."

"You say you actually knew Charles Darwin!" I blurted out. "That's impossible! Darwin lived and died more than a century ago, and he never visited America."

"Well, I'm not from America," the little bear insisted, "and I'm much older than you might think."

"How could you possibly be that old, and where exactly are you from?" I asked, wondering what outlandish claim I might hear next.

Darwin Bear did not disappoint. "Why, I'm from the Galápagos Archipelago, a group of islands located on the equator, about 600 miles (1,000 kilometers) west of Ecuador, in South America. And that's where I first met my friend Darwin, in October of 1835, when he was visiting the island that used to be my home."

My eyes practically popped out of my head when I heard this astonishing claim. It seemed utterly impossible to be talking with someone who knew Charles Darwin, who, along with Isaac Newton

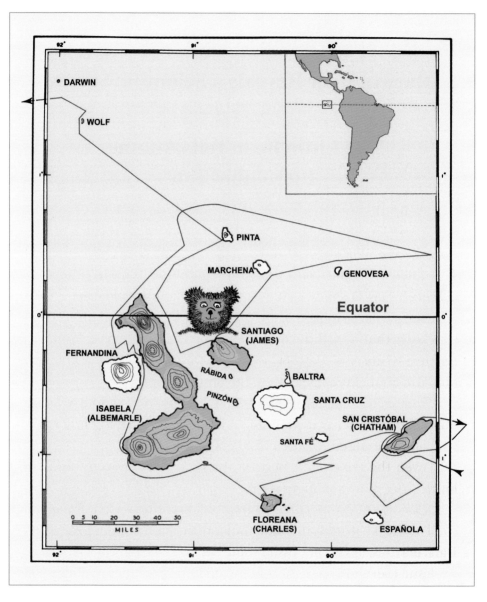

A map of the Galápagos Islands, showing Darwin's route on H.M.S. *Beagle* and the island (Santiago) where Darwin Bear claims to have made his home. Darwin visited the four islands shaded in green.

and Albert Einstein, is generally considered one of the most brilliant scientists who ever lived. More to the point, Darwin died in 1882, and if the two first met in 1835, that would make the bear upwards of a hundred and eighty years old. Unlike tortoises or redwood trees, bears generally do not live that long. Nor do they normally befriend genius scientists. So I had to be more than a bit skeptical.

"How do I know that you ever really met Charles Darwin?" I asked. "I don't mean to sound impolite, but you could be making up this whole story just to see how gullible I might be."

"Well," the little bear declared, "I know that Darwin's middle name was Robert, which is why his sisters sometimes called him Bobby; that he was the fifth of six children; that his *Beagle* voyage assistant's name was Syms Covington; that his various dogs were named Bobby, Button, Dicky, Pepper, Polly, and Tyke; that he liked to ride Galápagos tortoises; and finally that he loved gooseberries. I also know that he was born on the same day in history as Abraham Lincoln—February 12, 1809. Isn't that proof enough? Could someone who didn't know Darwin very well know all those details?"

"That's impressive," I admitted. "Very few people know such obscure facts about Darwin's life, except perhaps Darwin scholars."

"I know even more about Darwin," the little bear added, "things that even the most dedicated Darwin scholars could not possibly know."

"Like what?" I scoffed, not believing him at all.

The little bear shot back: "For example, I know exactly how Darwin reached his theory of evolution."

"But everybody knows that," I said. "Darwin was inspired to develop his theory of evolution when he realized that some of the islands in the Galápagos Archipelago were inhabited by their own form of mockingbird, giant land tortoise, and finch. This remarkable fact led him to conclude that animals and plants separated by geographic barriers, such as mountain ranges or oceans, eventually begin to evolve along separate pathways. Over time, diverging populations became separate breeds or geographic races, like the many different kinds of dogs and cats. Eventually, many of these populations evolved into new species altogether, producing the enormous variety of creatures that live on the earth today.

"This is also why animals and plants often tend to differ in striking ways from one continent to another. For example, Australia is the only continent that has kangaroos and platypuses, and South America is the only place where llamas live. Although geologists tell us that these continents were once united, that was several hundred million years ago. After the continents drifted apart, their inhabitants evolved in different ways."

The little bear had been listening patiently, but I could see that he wasn't particularly impressed.

"You seem to know the textbook account of how Darwin came to develop his theory of evolution," the little bear snorted, "but that's only part of the story. You have left out the most important part."

"And what 'most important' part is that?" I demanded.

"The part about bears, of course!" he said as he enthusiastically pounded the table with his paw.

"And what exactly do you mean by that?" I was sure that by pinning him down, I would catch my new acquaintance in another impossible claim. His story was becoming more and more unbelievable, which is often what happens with a series of little lies—they grow into ridiculously big ones. So I continued: "Scientists claim to have determined that there are no bears at all in the Galápagos Islands, and although there are bears in some parts of South America, Darwin never mentioned having seen any there or in any other place he visited during the *Beagle* voyage."

"That's what you and everyone else thinks," the little bear retorted, "but you're wrong, and I can prove it."

"How can you possibly prove such an astonishing claim?"

"Easily! I come from the Galápagos, and I'm a bear, so the Galápagos Islands *must* have been inhabited by bears! It's a simple

Buccaneer Cove on Santiago Island (also known as James), where Darwin camped for nine days. In the foreground on the right is a lava lizard. Lava lizards exist on some islands as distinctly different species. In the background on the left is a colony of land iguanas and their numerous burrows. In his *Journal of Researches* (1839)—the book he published about his experiences during the *Beagle* voyage—Darwin wrote, "I cannot give a more forcible proof of their numbers, than by stating, that when we were left at James Island, we could not for some time find a spot free from their burrows, on which to pitch our tent" (p. 469). Land iguanas have been driven to extinction on this island by the introduction of rats, pigs, and other animals.

A dome-shaped tortoise from Santiago Island, where
Darwin Bear came from.

matter of logic. I thought you were a scientist yourself. Aren't scientists supposed to be logical?"

Of course, what he had said was logical only if it were true, which
I seriously doubted. But bald-faced lies are hard to refute when the liar
keeps on insisting they're absolutely, positively, 100% true. I decided
the best thing to do was let the little bear ramble on with his story.

"Darwin camped on my island for nine days, collecting specimens of unknown birds, animals, and plants. I was minding my own
business, as bears tend to do, when Darwin spied me taking a nap
under a spiny *Acacia* tree. Not being sure what species I was, Darwin
tried to capture me in one of his big collecting nets. It's a good thing
we Galápagos bears are so nimble. His galumphing footsteps woke
me up and I quickly ran behind a large cactus. From there I yelled,

An Española tortoise, with its carapace turned up like a Spanish saddle. (The Galápagos Islands take their name from an old Spanish word for "tortoise.")

'Hey, Mr. Presumptuous Busybody, why are you trying to trap me? I haven't done you any harm. Besides, you're a guest on *my* island, and you should really be ashamed of yourself for even thinking of capturing your host. Otherwise, who is going to show you around the island and keep you from getting into trouble?'

"With a flustered look on his face, Darwin at once dropped the net and offered me his most sincere apologies—which naturally I accepted. I told him that if he wanted to capture anything, he should trap all of the goats on the island. Goats, as you probably know, were brought to the Galápagos by visiting sailors, and they were eating the blueberries that only we bears should have been eating.

"Darwin immediately agreed that the goats were wreaking havoc in the Galápagos environment. United by our low opinion of goats,

Darwin and I shook hands. I then told him the very story about bears and the Galápagos that inspired him to write his most famous book."

The little bear sat back with a satisfied smile. But his story wasn't over. "You see," he continued, "every island in the Galápagos was once inhabited by a different kind of bear. Darwin knew nothing of this, and at first he was inclined to disbelieve me. Much like you right now! So I asked him, 'Haven't you noticed that the mockingbirds are a bit different from island to island?'

"Darwin thought for a moment and said, 'Well, yes, I had noticed they were a bit different, but I just assumed these birds were local varieties and that the observed differences from one island to next are relatively unimportant variations within the same species.' 'They are hardly unimportant,' I insisted. 'In fact, there are four distinct kinds of mockingbirds in my archipelago, and three of them are confined to their own islands. Also, the tortoises and even some of the finches differ from island to island.'

"Darwin didn't believe me. He continued to think that the island-to-island differences among the bears, tortoises, and finches weren't truly differences between distinct species but just differences between local varieties. What could I say? The differences were plain as day to us bears. Humans, however, may be more short-sighted."

"So how did you convince Darwin?" I asked, wondering if I, too, would ever be convinced by the little bear.

"He admitted I might have a point, but he worried that my claims involved potentially revolutionary facts that could upset everything we knew about biology! He really didn't want to believe me—the truth was so inconvenient—and he also confessed that up to this point in his visit to the Galápagos, he had failed to label most of his collections by island and that most of his specimens from each island were hopelessly mingled with the rest. Where, then, was the proof of all this? Do you know what he said to me next?"

Of course I had no idea, so I shook my head.

A saddleback Galápagos tortoise eating the fruit of an *Opuntia* cactus. These prickly pear cacti grow in the Galápagos as tall as 40-foot (13-meter) trees on the eight islands where tortoises also live. On the Galápagos Islands without tortoises and land iguanas, *Opuntia* cacti grow only as low bushes, as they do in other parts of the world. Tortoises and land iguanas are avid consumers of prickly pear cactus pads and fruits, which are important sources of scarce food and water. Over time, these cacti have evolved into trees to avoid being eaten by these animals. On the hottest and driest of these islands, the tortoises have evolved carapaces displaying a rise above their necks (a characteristic saddle shape), which allows them to stretch their necks higher to reach vegetation. These various relationships reflect textbook examples of "coevolution," or the mutual evolution of two or more species in response to the changes in one another.

A volcanic landscape on San Cristóbal (Chatham), the first island Darwin visited. Darwin explored this region extensively, and in his *Journal of Researches* (1845 edition) he wrote, "Nothing could be less inviting than the first appearance. . . . The entire surface of this part of the island seems to have been permeated, like a sieve, by the subterranean vapours: here and there the lava, whilst soft, has been blown into great bubbles; and in other parts, the tops of caverns similarly formed have fallen in, leaving circular pits with steep sides" (pp. 373–74).

"He said, 'It appears that I may have bungled the opportunity of a lifetime.' " The little bear roared with laughter.

"What did you say to one of the greatest scientists of all time when he told you that he had made such a colossal blunder?" I asked.

"By this time I had gotten to like Darwin." The little bear smiled. "He was really 'a very nice guy,' as all the tortoises used to say. And because the tortoises live so long and have known so many generations of Galápagos creatures, they are generally excellent judges of character.* So I told Darwin not to worry, that I might be able to help, and that was the beginning of a long collaboration that culminated, two decades later, in publication of the *Origin of Species*."

The little bear slurped his tea, clearly delighted to have finished his long, involved explanation. I sipped my own cup, not at all sure what to think. Was I closer to the truth or further away than ever? Had this bear really been the one to get Darwin to think seriously about natural mechanisms for the theory of evolution? Was I incredibly lucky to be having tea with him, or just incredibly silly?

---

* The extraordinary longevity of the Galápagos tortoise, believed to be at least a hundred and seventy years, has inspired a Galápagos folk legend. According to this legend, the tortoise can see into the hearts of human beings. People who come to these islands to enjoy their wonders and natural beauty are allowed to go in peace. But those who come to the Galápagos to lay waste to its natural resources are struck by the curse of the tortoise (Octavio Latorre, *The Curse of the Giant Tortoise: Tragedies, Crimes and Mysteries in the Galapagos Islands*, 3rd ed. [Quito, Ecuador: Artes Gráficas Señal, 1999]).

# PART TWO

# A Revolution Takes Shape

Darwin Bear enjoying the view of Pinnacle Rock and Sulivan Bay, as seen from the summit of Bartolomé Island (just off the coast of Santiago, which is visible in the background). Boasting one of the most beautiful vistas in the Galápagos Archipelago, this bay was a favorite swimming and picnicking spot for Darwin Bear and his many penguin friends. The picturesque bay and island are named after Lieutenant Bartholomew Sulivan, a *Beagle* officer who was one of Darwin's closest friends and surveyed this region of the Galápagos Islands in one of the *Beagle*'s surveying boats.

CHAPTER 3

# The Little Bear Gives Darwin
# a Tour of Santiago Island

——— or ———

# Darwin Learns the Rules
# of Tortoise Racing

"Just how were you able to help Charles Darwin in his scientific work?" I asked the little bear. The safest thing—and most interesting, I thought—would be to go along with his story and see what more I could discover.

My fluffy friend launched into the next chapter of his account: "I had to get a message to all the bears living on the other islands in the Galápagos, telling them to come to Santiago Island, where I was living and Darwin was camped. Some sea turtle friends kindly offered to carry the news to each of the islands and transport the bears to meet with us.

"While we waited for the other bears, I gave Darwin a tour of my own island. He asked a lot of questions, stopping to perform little experiments—like pulling the tails of land iguanas when they were engaged in digging their burrows, just to see what they would do—and jotting down notes for his scientific research.

"I've given you a sense of Darwin as a scientist, but he was more than that. He quickly made friends with all the animals on Santiago. We all enjoyed his sense of humor. The tortoises, especially, used to laugh at his jokes. And it takes a lot to amuse a tortoise!

"I may have been an invaluable tour guide for Darwin, but he sometimes noticed things about my island that even *I* hadn't known."

Darwin Bear introduces Darwin to two land iguanas and a tortoise during their tour of Santiago Island.

The little bear looked a bit embarrassed at his own ignorance. I still wasn't ready to believe everything he said, but I couldn't help but be charmed by his story and the way he told it, so I didn't interrupt him but instead nodded for him to continue—which of course he did.

"For example, Darwin pointed out that many of the large volcanic craters in the archipelago—especially those near the ocean— are worn down on their southern flanks. He explained to me that this happens because the ocean currents in the Galápagos

A Galápagos land iguana resting under a giant *Opuntia* cactus tree. These large lizards, which live only in the Galápagos Islands, feed on grasses, annual plants, berries, and occasional insects. They are particularly fond of *Opuntia* cactus pads and the fruits and flowers of this cactus. Land iguanas are found only on five of the sixteen major islands. Two of the populations (on Santa Fé and the northernmost volcano of Isabela) are distinct species, differing from the form residing elsewhere, which constitutes a third species. These biological differences illustrate the role that geographic isolation plays in producing new species, as new species cannot evolve if the populations are too interbred, with not enough genetic separation between them.

Only about 20 inches (50 cm) tall, the Galápagos penguin is the second smallest of the world's eighteen penguin species, which obey Bergmann's rule. According to this rule, the nearer that closely related species reside to the equator, the smaller they become, so that they may more readily give off heat. The Little Blue Penguin, which is the world's smallest, is found in southern Australia and New Zealand, and has a height of just 17 inches (43 cm).

Islands consistently flow from south to north. Over time, these currents, and the winds they kick up, have eroded the craters on the south side.

"You may wonder why any of this matters, but I can tell you, it was like a light suddenly going on for me." The bear grinned. "I finally understood why my favorite swimming places were in the calmer waters of small bays on the northern sides of the islands, where we bears liked to play with our good friends the penguins."

The little bear paused for a moment, moving his hands in a swimming motion. "During our tour, Darwin asked me many questions about the land and marine iguanas that live there. He wanted to know what they ate, whether they were found on all of the

A Galápagos marine iguana—the world's only seagoing lizard. An excellent swimmer that propels itself through the water with the serpentine motion of its body and flattened tail, this species feeds off algae growing in tidal zones. As Darwin discovered, marine iguanas can remain under water for more than an hour.

This unusual lizard's closest relative is the Galápagos land iguana. The land and marine forms of iguana appear to have evolved from a common land-based ancestor 10 to 20 million years ago. This ancestral form lived on one or more islands to the southeast of the current Galápagos group and colonized new islands to the northwest. The new islands were created by volcanic action as the tectonic plate bearing the archipelago slid southeast over a "hot spot." Those older islands have long since eroded and sunk below sea level.

islands, and whether the marine form of iguana was afraid of sharks—along with lots of other information. You can find all of my answers recorded in his journal." My newfound friend seemed particularly proud of this last bit. And if something I'd said had ended up in one of Darwin's books, I suppose I would be bragging too.

"Darwin was particularly eager to learn more about the giant land tortoises," Darwin Bear continued. By now we'd drunk a whole pot of tea and eaten biscuits, scones, and even some of those little sandwiches with the crusts cut off. I still didn't know whether to believe him or not, but either way his stories were way beyond just interesting.

"Darwin was also curious to know just how far the tortoises could travel in a day." The bear dabbed genteelly at his mouth with his napkin; he really was an unusual bear. "So I suggested a tortoise race. That way he could measure their speed. Darwin wanted to set up the race right away, but as it was Tuesday, the tortoises' day off, I told him we would have to wait until the next morning. Besides," Darwin Bear lowered his voice to a confidential whisper, "I first had to teach Darwin how to maintain his balance on the back of a moving tortoise, as well as to acquaint him with the Rules of Tortoise Racing, which took us the rest of the day. You may not know it, but tortoise racing was considered so important among us bears that many formal rules had been developed over the years. Although speed was always an essential factor, in the end the winner was decided by a combination of speed and skill on the racecourse. To define skill, we needed formal rules. There was not enough time to teach Darwin all of these rules, so I laid out just the first seven. They were the most important ones anyway. They were also mostly associated with the collecting of Galápagos blueberries—our favorite food. Nothing mattered more to us bears than blueberries.

"If you plan on ever doing any tortoise racing yourself, you had better take careful notes. Rules are rules, and you can't observe

Darwin and Darwin Bear engaging in their first tortoise race. Darwin later wrote in his *Journal of Researches* (1839): "I was always amused, when overtaking one of these great monsters as it was quietly pacing along, to see how suddenly, the instant I passed, it would draw in its head and legs, and uttering a deep hiss fall to the ground with a heavy sound, as if struck dead. I frequently got on their backs, and then, upon giving a few raps on the hinder part of the shell, they would rise up and walk away; —but I found it very difficult to keep my balance" (p. 465).

half a rule and get credit for the whole. Winning a tortoise race depends on knowing and obeying all the rules—each one to the letter."

I wasn't particularly planning on entering a tortoise race, but you never know when such information might turn out to be useful. So I did as Darwin Bear instructed and wrote down the seven rules.

## Seven Basic Rules of Tortoise Racing

RULE ONE: Picking a ripe blueberry along the racecourse equals 10 seconds of racing time, which is then subtracted from the racer's overall time. So if a bear riding on a Galápagos tortoise collects 6 ripe blueberries, he will have a whole minute deducted from his finishing time. This means that a bear who finishes 50 seconds sooner, but who has not collected any blueberries, would still lose the race by 10 seconds. By contrast, if a bear picks an unripe blueberry, a penalty of 20 seconds is added to the final time. Picking unripe blueberries merit this penalty because it is considered ecologically wasteful.

RULE TWO: Picking an especially large blueberry—defined as a blueberry that is larger than a full-grown bear's nostril—counts as a 20-second advantage in racing time, but only if the berry is ripe.

RULE THREE: All tortoise racing routes must go around blueberry bushes. Knocking into a blueberry bush triggers an automatic penalty of 10 seconds, which is added to the contestant's racing time. Breaking off a branch draws an even stiffer penalty of 20 seconds.

RULE FOUR: Scaring off a feral pig, donkey, or goat that is engaged in eating a blueberry bush is richly rewarded by the subtraction of 25 seconds from the contestant's racing time. Hitting one of these animals with any kind of object, whether or not they're eating a blueberry bush, earns the contestant 20 seconds in reduced racing time, and a total of 45 seconds if the animal also runs away.

# OFFICIAL TORTOISE RACING RESULTS
### (for the Three Contestants Participating on 10 October 1835)

| Racing Event | Darwin Bear | Darwin | Covington |
|---|---|---|---|
| Total racing time (unadjusted) | 25:10 (3rd place) | 25:03 (2nd place) | 24:10 (1st place) |
| Ripe blueberries captured (with an adjusted racing time) | 25 (21:00) | 20 (21:43) | 10 (22:30) |
| Extra-large blueberries captured (with an adjusted racing time) | 10 (17:40) | 11 (18:03) | 2 (21:50) |
| Bumped blueberry bushes (with an adjusted racing time) | 0 (17:40) | 0 (18:03) | 1 (22:00) |
| Broken blueberry branches (with an adjusted racing time) | 0 (17:40) | 0 (18:03) | 3 (23:00) |
| Wild animals chased away from blueberry bushes (with an adjusted racing time) | 4 (16:00) | 1 (17:38) | 0 (23:00) |
| Wild animals hit with blueberries or other objects (with an adjusted racing time) | 3 (15:00) | 1 (17:18) | 0 (23:00) |
| **Total *adjusted* racing time, based on the Rules of Tortoise Racing** | **15:00** **(1st place)** | **17:18** **(2nd place)** | **23:00** **(3rd place)** |

Leading in one of their tortoise races, Darwin Bear prepares to pick two extra-large and fully ripe blueberries, worth 40 seconds of racing time. Behind him, in second place, an alert Charles Darwin is attempting to scare two wild goats away from a blueberry bush by throwing one of his largest blueberries at them—thereby sacrificing 20 seconds of racing time in the hope of achieving a 50-second net reduction in his overall time. Trailing Darwin, in third place, is Syms Covington, who, as Darwin's loyal research assistant, had stopped earlier to capture a grasshopper.

**Rule Five:** There are no tortoise races between 10 a.m. and 2 p.m., because it is usually too hot, except in the highlands, where racing can safely proceed all day. (For a definition of where the "lowlands" end and where the "highlands" begin in the Galápagos, consult Rules 312–15 in the 784-page *Rules and Regulations for Tortoise Racing in the Galápagos*—see Appendix 2, pp. 135–36.)

**Rule Six:** There are no tortoise races on Tuesdays, because this is the tortoises' day off.

**Rule Seven:** There are no tortoise races on days when the bears decide there will be no tortoise races—which happens whenever the bears are feeling tired, filled with too many blueberries, conducting blueberry plant censuses, planting new blueberry bushes, entertaining visitors, planning or engaging in picnics, or otherwise preoccupied.

When I had finished recording the seven rules that Darwin Bear had described to me, he launched back into his story.

"Darwin was very pleased to be able to determine, as a result of all our tortoise racing, just how far a tortoise can travel in a day. As he later reported in his *Journal of Researches,* the average tortoise travels about 360 yards an hour, resulting in a total distance traveled, each day, of about 4 miles—allowing, as Darwin noted, 'a little time for it to eat on the road.' Regarding the tortoises' diet, which Darwin also studied carefully, he noted that they particularly liked the berries of the endemic Galápagos guava, which we call 'guayabillo.' Darwin himself described these guayabillo berries as being rather 'acid and austere.' To us bears, they tasted much like turpentine, but tortoises, you know, don't have very discerning palates.

"Long ago, the most experienced tortoise racers discovered that their tortoises usually performed better during races if they were given some guayabillo berries just before the race began. Of course, I didn't tell Darwin this, which is one of the reasons I generally bested him and his assistant Syms Covington in our many tortoise races. As we bears used to say, 'All's fair in love and tortoise racing, if you really want to be a champion tortoise racer.' "

A map of the places Darwin visited on Santiago between October 8 and 17, 1835. Also shown are the locations where two of the longer tortoise races took place between Darwin Bear, Darwin, and Syms Covington. The first tortoise race was from the salt mine crater to the coast. Another, longer race was from Darwin's campsite at Buccaneer Cove to the summit of the island. This second race was so lengthy that it had to be divided into three segments, to give the tortoises time to rest. Although Covington in the first race, and Darwin in the second race, had the fastest times, Darwin Bear won both races based on points scored according to "The Rules of Tortoise Racing."

Having carried a message from Darwin Bear to the other islands, three sea turtle messengers return with fifteen bears.

## CHAPTER 4

# The Little Bear Discloses
# What He Told Darwin

— or —

# Startling Facts Come to Light
# and a Crisis Is Averted

The little bear paused for a moment, apparently lost in fond memories of days spent in tortoise races with Darwin. As preposterous as his story sounded, a part of me couldn't help believing him; it seemed that he really had known Darwin, and this made me want to strengthen our acquaintance. So I suggested that we go to my house to continue our conversation, and he accepted my invitation.

Having arrived at my home, we settled into the living room. I made more tea and put out some crackers, biscuits, and little sandwiches, as the bear still seemed to be hungry. "Do go on," I urged him.

Darwin Bear patted his tummy. "Thank you, I haven't eaten in a long time."

He then resumed his story. "One day after a long walk, Charles and I returned to our camp in Buccaneer Cove to find that the bears from the other islands had arrived, carried on the backs of sea turtles. Altogether there were fifteen bears—one from each of the other major islands. There were big bears, little bears, medium-sized bears, brown bears, tan bears, tawny-colored bears, and even a tiny bear carrying a miniature tennis racket carved from a bird bone.

"The bear from Fernandina Island arrived with the fur on his backside singed. A volcano was erupting when the sea turtles arrived at this island. This Fernandina bear, who was the last of his kind on the island, had narrowly escaped being swallowed up by smoldering lava flows as he desperately fled toward the ocean. Fortunately, the sea turtles found him bobbing about in the sea, just offshore at Punta Espinosa.

"Because the bears had not seen one another since the last Tour de Galápagos—a seven-day, multistage race held once a year—they had some catching up to do. A girl bear from Pinta Island, where land tortoises had been all but exterminated by buccaneers and whalers, reported having been chased and nearly caught by a party of visiting whalers. She had managed to escape only when two very large tortoises blocked her pursuers' way. Sadly, the whalers grabbed the two tortoises instead, along with more than a dozen others, to be turned into turtle soup.

"Another of the girl bears, from Marchena Island, had much more pleasant news to report: namely, that a longstanding feud between the sea lions and the marine iguanas had finally been settled. The sea lions had long enjoyed grabbing the tails of the marine iguanas and tossing the helpless animals back and forth in a game of iguana water polo—a fun sport for the sea lions, but not for the iguanas. After a lengthy arbitration presided over by the Marchena bear, the sea lions agreed to limit their iguana polo sessions to once a month. The iguanas, in exchange for their monthly participation, would be given directions to any rich growths of algae the sea lions found while hunting for fish.

"Darwin was astonished at the diversity among us bears, and once we had finished trading stories, he peppered us with questions about it. 'Do you really mean,' he asked, 'that *every one* of these bears is from a different island, and that on each of these islands only one kind of bear is ever found?' I told him that was precisely the point I'd been trying to make, if only he would listen.

"I also described how we Galápagos bears lived almost exclusively on blueberries. For millions of years, we had adapted our-

The new arrivals, along with two tortoises and their good friend El Pingüino, introduce themselves to Darwin and begin telling him about the many mysteries of their remarkable islands.

selves to blueberry consumption on each island. On large, tall islands like mine, where there is much *garúa* (a drizzly mist), the blueberries were large and plump, and bears, too, generally had big bodies and large tummies in order to digest them. On smaller, drier islands, where blueberries were less plentiful, the bears were

Tennis Bear uses his racket to get blueberries.

correspondingly smaller, with smaller tummies. On the smallest of the bears' islands, the bears were so tiny vis-à-vis the blueberries that they needed to use little tennis rackets to dislodge them from the blueberry bushes. This explains the tennis-racket-carrying bear.

"Being clever creatures, we had developed a plausible theory to explain these island-to-island differences. I'll tell you now what I told Darwin. We had long noticed that on each island, mother bears were always looking for the biggest blueberries to give to their cubs. We called this process by which mother bears helped their cubs to become big and chubby 'maternal berry selection and survival of the fattest.' On the small, dry islands, there simply were not enough berries for bears to become big and fat. Tennis bear mothers, for example, could usually find only a few pawfuls of berries each day, and this harvest was barely enough to fill their own tummies and those of their baby tennis bears. Hence the development of these particularly small-tummied bears was what we called 'maternal berry selection and survival of the thinnest.'

"Similarly, the best tennis-playing bears were often the best at gathering blueberries because a really strong and accurate swing of the racket was needed to dislodge enough blueberries for adequate daily consumption. Naturally, the bears that were most adept at tennis were also the most successful at raising healthy and talented offspring like themselves, as tennis skills tended to run in the family. Their descendants then passed on their own superior athletic abilities to the next generation, and this process was repeated,

After teaching a baby Galápagos tortoise to play tennis, Tennis Bear easily wins game, set, and match against his eager but slow-moving opponent. On the right, a land iguana referees the match.

generation after generation, until all the tennis bears became highly skilled at playing tennis as well as collecting blueberries.

"After Darwin heard me talk about this process of development, his eyes lit up and he said, 'If you bears could be shown to be separate species, rather than just varieties, the information you have provided would undermine one of the most crucial beliefs in science—namely, that species are fixed for all time and hence immutable. Such new evidence would revolutionize the way we think about life on earth. In fact, this remarkable evidence might even change the way people think about themselves and their place in nature!'

"I could not let this remarkable statement pass without asking Darwin just what he had in mind. He replied by saying, 'The exquisite adaptations we so frequently find in nature have long been attributed to a designer, or a clever cosmic watchmaker, as William Paley liked to argue in his widely read book *Natural Theology, or, Evidences of the Existence and Attributes of the Deity*. I have always found Paley's argument to be compelling. As Paley explained, when we come upon an elegant adaptation in the natural world, like the plumes that allow some seeds to waft through the air or the interlocking hinges of a bivalve shell, it is much the same as if one were walking along and suddenly found a watch lying on the ground. Surely, this watch must have had a watchmaker, just as there must have been an intelligent designer of all of the marvelous adaptations exhibited by different forms of life. But, in contrast to Paley, you are suggesting that the Galápagos bears actually evolved on their own, without the need for such a wise designer.' "

I had been patient during the description of the tortoise races, the blueberry picking, and the "survival of the fattest" theory, but I could keep silent no longer. "This whole story about bears and blueberries seems increasingly farfetched. I'm too polite to have contradicted you when you were telling me about the important role that blueberries played in the rules of tortoise racing, but I have to say now that, to my knowledge, there are no native blueberries in the Galápagos Islands."

"That's *precisely* my point!" the bear retorted. "For if there were any blueberries in the Galápagos, then you would know that I was *not* telling the truth.

"You see," he continued without skipping a beat, "we bears became so good at blueberry collecting that our blueberry supplies on each island had begun to dwindle. Over time, we were able to

Riding on the back of a baby Galápagos tortoise, Tennis Bear finds some
of the last blueberries on his island and reaches to swat them down with
his racket.

adapt successfully by becoming smaller, which meant that fewer blueberries were required to sustain us through the rigors of blueberry
hunting over generally difficult terrain. That's why Galápagos bears
are noticeably smaller than the average bear.

"Then visiting buccaneers and whalers introduced rats, goats,
pigs, donkeys, and other domesticated animals to the islands,
and these intruders began to eat everything in sight, including,
of course, our beloved blueberries. At first, we tried to ration the
blueberries and hide them from the goats and pigs—who refused
to follow our own efforts at blueberry conservation. One by one, the
plants kept disappearing. Bear populations on islands like mine—
islands that had once supported thousands of bears—were gradually

decimated, leaving only those bears that were especially adept at blueberry collection. Eventually there were only enough plants to support a single bear on each island, leaving just sixteen bears in the archipelago as a whole.

"You can imagine how depressing all this was, as well as how disruptive to our tortoise races and group picnics. Occasionally we would all get together on one of the sixteen main islands to conduct a tortoise race or have a nice picnic, but most of the time we just lived alone, each on a different island, searching for whatever blueberries remained. This was the sad and precarious state we bears had reached when Darwin arrived in the Galápagos."

"So there are no blueberries *now*," I exclaimed, "but at one time there were? Come on, their absence is hardly proof of their having once existed!"

"Well, fortunately what you choose to believe about my story is not all that important," Darwin Bear shot back. "What is important is that Darwin believed us and that he saw clearly the only solution to our predicament. He told us that if he left us on the islands we would eventually starve, so he offered to take us all back to England with him on the *Beagle*. He promised us lots of delicious berries in England, and he even invited us to live with him. He further suggested that we bring any remaining seeds of our native blueberries with us, so that we could cultivate them in England in his garden. You see, he had grown rather fond of us, which is certainly understandable. Who wouldn't?

"What would *you* have done?" Darwin Bear asked me, before quickly answering his own question. "The idea of plentiful berries, including our beloved blueberries, was just too tempting! So we decided to accept Darwin's offer. Besides, we were all very curious to know how other kinds of berries tasted—particularly Darwin's favorite, gooseberries.

"We held a farewell party to say goodbye to all of our Galápagos friends—the tortoises, iguanas, penguins, and finches, as well as all

The Galápagos bears heading to England on H.M.S. *Beagle*.

of the other birds and reptiles—and we scrambled onto the *Beagle* with our new friend Charles. Naturally, we brought with us the last of the Galápagos blueberries, and this is why there are no longer any blueberries, or bears, in the Galápagos Islands today. Of course, we took special care not to eat all of our remaining blueberries so that we would have a supply of seeds with which to grow our native Galápagos blueberries in England.

"As the *Beagle* sailed away, we bears were so relieved and happy that we composed a song to commemorate our salvation by Darwin. Would you like to hear it?" And, without waiting for an answer, Darwin Bear began to sing:

## THE BEARS' SALVATION SONG
### (Sung to the tune of "God Save the King" [or "Queen"] or "My Country 'Tis of Thee")

1. We bears love Charlie D.,
   It comes so naturally,
   He saved us bears.

   We've had the greatest thrills
   With Darwin in the hills,
   He knows all the rocks and rills,
   And he saved us bears!

2. With berries everywhere,
   Bears didn't have a care,
   They ate their fill.

   Then came that awful day,
   Berries had gone away,
   All bears could do was pray,
   "Who will save us bears?"

3. The birds were sad to see
   Bears in such misery,
   "Please save the bears!"

   Tortoises offered rides,
   Sea turtles wiped their eyes,
   But to no real surprise
   Nothing helped the bears.

4. Our Charlie had a plan,
   He was so wise a man,
   He loved us bears.

In England berries grow
In places I can show,
Bears there will beam and glow,
Darwin saved us bears!

5. On board the *Beagle* now
Bears gathered at the bow,
To see the sights.

Bears love to sail the seas
And feel the tropic breeze,
Soon England's shores we'll see,
Darwin saved . . . us . . . *bears*!

(Composed by the Galápagos bears, aboard H.M.S. *Beagle*, in honor of their friend Charles Darwin for saving them from certain extinction.)

A Galápagos Finch-Bear, a mix of bear and finch. Because the hybrid finch-bear was able to harvest so many blueberries, the fruit became scarce and then finally vanished. Without its main source of food, the Finch-Bear also became extinct.

## CHAPTER 5

# How Darwin Bear Got His Name and Helped Develop a Revolutionary Theory

#### —— or ——

# A Widespread Scientific Legend Is Dead Wrong

"Once we were all aboard the *Beagle*," Darwin Bear continued, "we realized we had a problem. We were so accustomed to living as solitary bears on our separate islands that we hadn't needed names for quite some time. In fact, it had been so long since we had last used names that we had all forgotten whether or not we ever *had* any names.

"At first, we just called one another by the names of the islands on which we had lived, but we soon agreed that these names were too impersonal. So we decided to give ourselves new names. Because some of the bears wished to try out their new names for a while to see whether they were a good fit, this naming process was not fully completed until after we had reached England and gotten ourselves settled into our new home in London.

"We all agreed that at least one of us should be named in honor of our new friend Charles. Because I had met Darwin first, the other bears proposed that I should be named Darwin Bear. I was, and still am, called Darwin *Bear* rather than Darwin to avoid any confusion with 'the big guy.'

"Darwin helped us pick most of the other names, suggesting those of his closest friends and other people he particularly admired. So, among the other boy bears, there was a FitzRoy Bear, named

The Galápagos bears lining up to receive their new names. Darwin Bear is announcing the names, and Covington Bear is recording them in the *Beagle*'s logbook.

after the captain of the *Beagle*—because this particular bear, like Captain Robert FitzRoy, loved to sail; a Lyell Bear, named after the geologist whose insightful theories guided Darwin throughout the *Beagle* voyage; a Henslow Bear, named for Darwin's teacher, who secured his appointment as ship's naturalist aboard H.M.S. *Beagle*;

and a Karl Ernst von Bear, named for the famous German embryologist whose research provided an important contribution to the *Origin of Species.*

"One of the girl bears who liked to stargaze became Caroline Herschel Bear, after the astronomer who catalogued nearly twenty-five hundred nebulae and also discovered several new comets. Darwin was particularly pleased with this name choice, given his admiration for her equally brilliant nephew, John Herschel. As Darwin later asserted in his *Autobiography*, his reading of John Herschel's *Preliminary Discourse on the Study of Natural Philosophy* when he was still a Cambridge University undergraduate had inspired in him 'a burning zeal to add even the most humble contribution to the noble structure of Natural Science.'

"Another of the girl bears became Emma Darwin Bear, named after Darwin's cousin and later his devoted wife, who helped him to revise his manuscript drafts and to proofread the galleys of his books. A third girl bear became Martineau Bear, adopting the last name of the famous social theorist who was a close friend of Darwin's brother, Erasmus. Because of her sometimes controversial writings about society, Harriet Martineau has often been cited as the first female sociologist, and Darwin personally considered her a fascinating thinker.

"Like her namesake, Martineau Bear was distinguished by her clever writing in pamphlets circulated among the bears. She was particularly known for having promoted social reforms among the bears, including—in the days when each island still had many bears—a social-security system for supporting old or injured bears when they could no longer hunt for blueberries on their own. Additionally, it was Martineau Bear who had settled the dispute between the sea lions and the marine iguanas when she was still residing on Marchena Island.

"The tiny bear with the tennis racket was first named after Lieutenant Bartholomew James Sulivan, one of Darwin's closest friends

on the *Beagle*. This bear tried out his new name for a month or so
before deciding that he didn't like it because it was too long—so
much bigger than he was. So Darwin suggested Stokes, another of
his *Beagle* friends. But this name was too short. Because he was much
tinier than all the other bears, the bear wanted a name that wasn't
too big or too small and would also make him feel special. Darwin
suggested several other distinguished names—those of several
British prime ministers and a series of famous scientists, including
Galileo and Isaac Newton. Over the next several weeks, the tiny bear
tried out each of these names, but in the end none felt right. Finally,
the other bears began calling their smallest member Tennis Bear,
and the name stuck.

"Back in England, four of the bears later changed their names
after new and particularly admired people came into Darwin's life.
But I don't want to bore you with all these naming details, as I have
much more important things to tell you."*

"What did you do with yourselves," I asked, "after you were settled in England? You must have had a lot of free time, since Darwin
kept you all supplied with plenty of tasty berries and you no longer
had to hunt for them."

"We immediately set to work to help Darwin write his book about
the voyage of the *Beagle*. We also helped Darwin resolve a number
of crucial scientific problems. It was FitzRoy Bear, for example, who
first told Darwin that he believed whales might be descended from
ocean-going bears. We were a very effective research team! Without
us, could Darwin have published sixteen major works? Don't you
think that's a remarkable coincidence?"

"What do you mean by 'remarkable coincidence'?" I asked.

"Why, the coincidence that we are sixteen bears and that Darwin

---

* Darwin Bear later gave me a complete list of all of the bears' names, together with
information about their special talents, islands of residence, and personalities. This
information is reproduced in Appendix 1 (pp. 127–33). Darwin Bear returns to the
issue of name changes in the following chapter.

The Galápagos bears, helping Darwin to research and write his various books.

published sixteen different studies, if we count his geological results from the *Beagle* voyage and his subsequent work on barnacles as being two of these sixteen research projects. Of course, it's *not* a coincidence! A different bear helped Darwin write each study—which partly explains his brilliant career.

"One of our greatest contributions to Darwin's research, which brought about his final conversion to the theory of evolution, involved our part in proving that the islands in the Galápagos really were each inhabited by their own particular species, and not just varieties. The ornithologist John Gould also played a leading role here by convincing Darwin that at least three distinct species of mockingbirds were confined to different islands, and that two species of vermilion flycatchers and perhaps some of the finches were also restricted to their own islands.

"Just as important, Gould persuaded Darwin that the now famous Galápagos finches were not, as Darwin had initially thought, members of four different families of birds—wrens, blackbirds, grosbeaks, and finches. Instead, Gould placed all of these birds in a single closely related subgroup of finches. This correct conclusion helped Darwin realize that all those diverse finches must have evolved over time to fill the various empty niches that were available in the Galápagos when animals and plants first began to colonize the archipelago several million years ago.

"Furthermore, because the islands vary in height, local climate, and resident species, Darwin concluded that evolution had consistently favored different attributes in the animals and plants of each island. And because these island populations were geographically isolated from one another, the accumulated changes led to the development of separate races and eventually new species. That's evolution in a nutshell, right? Those species having the traits best suited for each place thrived and passed these traits on to their offspring, who in turn passed them on to theirs, and so on for many, many generations.

"Although Darwin had great admiration for Gould's skills as an ornithologist—that means bird scientist, in case you didn't know—he also recognized that some of Gould's taxonomic classifications of specimens into families, genera, and species might be subject to revision. Other ornithologists could, and occasionally did, disagree with how Gould had classified these birds, and Darwin realized that this element of doubt could cause people to question his evidence for evolution through geographic isolation.

"Reinforcing Darwin's concern was the fact that Gould was known as a taxonomic 'splitter' in his identifications of new species. This tendency toward mistakenly splitting a single valid species into two or more specific forms was often reinforced by the scientific custom of attaching the name of the first describer to the name of the new species. The Floreana mockingbird, for example, is now known as *Mimus trifasciatus* ( J. Gould), because Gould was the first naturalist to name this novel taxonomic form, along with two other of the four currently recognized species of Galápagos mockingbirds. Gould's name is also attached to eight of the seventeen species of Darwin's finches.

"Darwin thought this practice of self-attribution was an impediment to good science, and we all soon found ourselves facing an unexpected and up-close instance of this problem. When Gould Bear learned that his namesake got to attach his name to every new species he was the first to identify, he wanted to be famous too, by naming things he saw in the neighborhood of our new home. He would say things like 'Look over there on the ground—there's a large, brown worm.' Then he would quickly consult the Latin dictionary he always carried with him to provide proper Linnean names, after which he would declare, 'I am going to call this worm *Vermis brunneis* (Gould Bear).' Once, he even tried naming a new crescent moon he was the first to spot, blurting out, 'Everyone, there's a *Nova luna* (Gould Bear).'

"Fortunately, Gould Bear eventually gave up this annoying habit after we all started making fun of him, naming all kinds of silly things ourselves, from a juicy gooseberry (*Berrius fattus*) to a new shirt (*Shirtalus buttonus*). Still, in his letters, he always signed himself as Gould Bear (C. Darwin), since Darwin had been the first to name him. One very good thing, however, did come out of Gould Bear's 'name-everything-first' phase. He ended up learning quite a bit of Latin, and this later proved useful to Darwin in his books on living and fossil barnacles when he needed to come up with Latin names for hundreds of new species.

"But I've gotten off track," Darwin Bear continued. "We were talking about John Gould."

I hadn't been talking about anyone. I was just listening. And I was still trying to decide whether this bear was the most inventive liar I'd ever heard or a phenomenal creature who had indeed lived

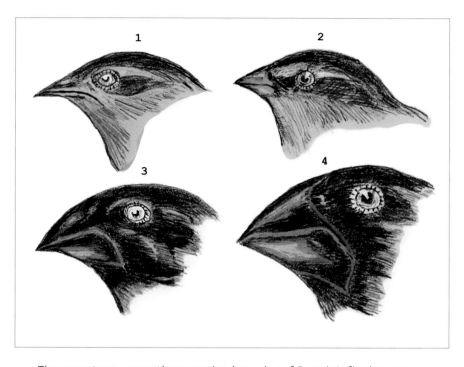

The seventeen currently recognized species of Darwin's finches represent a classic case of adaptive radiation. The first ancestral colonist in
the Galápagos arrived more than a million years ago and was similar in
form to the present-day Warbler Finch (1). This first colonist was not a
true finch but rather belonged to the tanager family (*Thraupidae*). The
Small Tree Finch (2) and eight other tree-finch species were part of a
later evolutionary branch giving rise to additional insectivorous species. The Medium Ground Finch (3), Large Ground Finch (4), and six other
ground finches, which all forage largely on seeds, were the last of the
seventeen species to evolve, beginning around 300,000 years ago. (After
a woodcut in the second edition of Darwin's *Journal of Researches* [1845,
p. 379])

with Charles Darwin and influenced his theories. Preposterous! And
yet. . . .

Darwin Bear took up his story. "As I was saying, Gould had made
a good case for the Galápagos mockingbirds, but Darwin still had no
other solid proof of this evolutionary process, because he had failed

to label by island most of his Galápagos birds and other animals. He had kept his plant collections separate by island, however, because these specimens needed to be flattened and dried immediately in plant presses, and this fortunately prevented them from being mingled. It took another ten years for Darwin's friend Joseph Hooker to make a careful taxonomic analysis of these botanical species, which overwhelmingly confirmed the conclusion that each island in the Galápagos tends to have its own endemic forms. When Darwin finally heard about these results, he wrote to Hooker to say how glad he was about the findings and 'how wonderfully they support my assertion on the differences in the animals of the different islands, about which I have always been fearful.'

"In fact, Darwin was so grateful to have the compelling evidence provided by Hooker that he suggested that one of the bears might want to consider becoming Hooker Bear. Hooker himself had previously complained to Darwin, saying 'I'm your best friend. Why isn't one of the bears named after me?' The bears, however, weren't so sure, as they were all quite happy with their current names.

"Undeterred, Hooker kept campaigning for the name change, in part by sending Darwin and the bears abundant supplies of bananas grown at Kew Gardens, where he was the director. The bears just loved those bananas, which the Darwin family nicknamed Kew gooseberries. The bears always looked forward to Hooker's visits to Darwin's new home, about 15 miles south of London in the Kentish village of Downe, to which Darwin had moved in 1842. The bears could be certain that, with each Hooker visit, there was sure to be a delicious Kew gooseberry feast. At last, Wickham Bear announced in 1850 that he was going to become Hooker Bear. Because Wickham Bear gained quite a bit of weight just before he announced this name change, the other bears, and Darwin too, strongly suspected that Hooker must have bribed Wickham Bear with extra Kew gooseberries.

"Hooker greatly endeared himself to the bears because of their shared love of the many gooseberry varieties grown in Darwin's garden. It was a distinct source of family humor that Hooker was always seeking to wangle an invitation to visit the Darwins during the height of the gooseberry growing season. In 1865 he even wrote to Darwin to say that if Darwin's daughter Henrietta 'lets the Gooseberry season go bye without inviting me I will kill her.' This strong mutual love of gooseberries also created a special bond between Hooker and Darwin's children; they, too, were devoted to gooseberries and somewhat astonished that Hooker obtained such delight from eating them in the family's kitchen garden together with the children.

"Darwin's growing post-voyage conviction that all species had likely evolved was fortunately strengthened by Richard Owen's brilliant study of Darwin's vertebrate fossils. These South American fossils—which included giant ground sloths, giant armadillos, and a large, llama-like creature—clearly supported what was called the 'law of succession': that fossil forms on each continent are almost always most closely related to the creatures living there today rather than to the species found on other continents. Still, Darwin was always on the lookout for additional compelling evidence."

I finally dared to interrupt. "Why didn't Darwin just use you bears as his evidence? Surely you had provided him with a textbook case of evolution in action—as good as, or even better than, that of the mockingbirds, tortoises, and the now famous Darwin finches."

"I thought you were a scientist yourself," the little bear muttered. "You should know that science rests upon the possibility of other scientists being able to confirm and replicate one another's work. Since we bears were the *last* of all the Galápagos bears, there was no possibility that other explorers could return to the Galápagos and validate Darwin's claims about us. Look how skeptical *you* were about my story, given the absence today of both bears and blueberries in the Galápagos. However, we bears were useful to Darwin in another very important way—we found a solution to his problem!"

"Just how did you do that?" I asked, leaning back and wondering what colossal lie I was going to hear next. The funny thing was, I was

Darwin Bear writes a letter to Captain Robert FitzRoy, asking for help with the missing scientific evidence from the Galápagos Islands.

actually beginning to believe some of these extraordinary claims—or at least the bear's lies did not seem quite so preposterous anymore. What kind of scientist did that make me, I wondered, to accept such unlikely testimony without a shred of proof?

"You see," Darwin Bear explained, "although Darwin had failed to label his collections by island, the situation was not entirely hopeless. While chatting with some of the crew members on the *Beagle*, we had learned that the captain and two of the crew had also made collections of birds during their Galápagos visit, and they had a written record of these localities. So we suggested that Darwin borrow those specimens and use them as evidence to support his theory of evolution. Darwin was so excited by our proposed solution to his problem that he rewarded us with extra blueberries!"

"You mean to say that the crucial evidence Darwin used to support his famous theory of evolution by natural selection was originally *borrowed* from someone else?" This, I thought, was unquestionably a fib.

"Indeed," Darwin Bear nodded. "Science is much more of a social activity than people think. Researchers sometimes use the ideas and findings of other researchers, just as bears often share their blueberries with one another and even reveal the location of the most abundant blueberry bushes. Science is not just facts and figures, you know. Researchers both compete and cooperate with one another at different times. As Darwin understood, these social interactions help to determine which ideas are ultimately successful. You might even say that ideas themselves evolve in a Darwinian manner, by survival of the fittest theories. But this is another topic, and I really shouldn't stray again from my story, especially since I haven't gotten to the most important part."

"And just what most important part do you have in mind?" I asked.

"Can't you guess?" he replied.

# PART THREE

# Darwin's Big Idea

The long, pointed bill of the Cactus Finch, one of the eight ground-finch species, is specialized for feeding on *Opuntia* cactus. The flowers, fruits, and seeds are a major food source for this finch.

# CHAPTER 6

# How Darwin Hit on His Big Idea

—— or ——

# Darwinian Theory Triumphs with the Help of the Bears

"I am trying to be modest here," Darwin Bear continued, "but that's not easy, as we bears contributed to Darwin's work in so many different ways. Sometimes we would recommend books and articles for Darwin to read. For example, in September 1838, Martineau Bear suggested that Darwin read the final edition of Thomas Malthus's *Essay on the Principle of Population*, which she had borrowed from Darwin's brother, Erasmus, who in turn had taken an interest in the book after Harriet Martineau brought it to his attention. Malthus's famous book argued that populations can increase at astonishing geometrical rates as long as they are not checked in their growth by naturally occurring impediments, such as famine or disease. By contrast, the available food supply in nature is relatively constant, and even among human populations, Malthus argued, it could increase only at an arithmetic rate. This means that the food supply would always be a relentless check on population growth.

"This insight inspired Darwin to develop the final step in his theory of evolution by natural selection. As he was reading Malthus's book, Darwin suddenly realized that what the bears had told him about their own evolution was not an exception associated with the bears' heavy reliance on blueberries, but rather the rule. Population pressure, Darwin now saw, provides an ever present agent of

evolutionary change. Any heritable variations in an organism—such as Tennis Bear's sporty reflexes—would be favored in the everyday Malthusian struggle for existence. These variations would then be passed on to offspring, providing a mechanism for endless evolutionary changes of an adaptive nature.

"Darwin was now able to extend our own theory about 'maternal selection and the survival of the fattest' to include *all* animals and plants, using his new and broader concept of 'natural selection' and survival of best-adapted organisms. Over time, this gradual process of evolutionary change was what had caused some seeds to develop plumes that distributed them through the air, and Galápagos finches to evolve differing beaks to be more efficient at finding food.

"As Darwin also recognized," Darwin Bear continued, "the same refining process over many generations explained why the different islands in the Galápagos group were inhabited by different species of land iguanas, tortoises, birds, and plants. 'At last,' Darwin announced in the middle of his reading of Malthus, 'I have a theory by which to work.'

"At this time, Darwin began to be bothered by a number of moral and religious questions. Led by Martineau Bear, who understood these issues best, we tried to help him resolve some of the thornier concerns. The biggest one, of course, was that his theory contradicted religious teachings about the creation of the world. According to some theologians, this Creation had taken place on October 23rd, 4004 BC—at about nine o'clock in the morning, to be precise. People like Bishop Samuel Wilberforce, who considered himself a direct descendant of Adam and Eve and not some hairy ape-man, couldn't accept evolution. Furthermore, Darwin's own father had wanted him to become a clergyman, and many of Darwin's friends, including his revered teacher John Stevens Henslow, had strong religious convictions that kept them from accepting this theory.

"Martineau Bear helped clarify all this when she asked Darwin how many religions exist in the world. He admitted that there must be hundreds. 'Besides Christians,' he responded, 'there are Buddhists, Hindus, Muslims, Jews, as well as the many beliefs held by local tribal groups in remote parts of South America, Africa, and Australia.'

" 'And do all of these different religions claim that species cannot change? Or that the world was created in 4004 BC?,' Martineau Bear pressed.

" 'Of course not,' Darwin replied. 'I was thinking only of Christianity, as written in the Bible.'

"That gave Martineau Bear a chance to point out the obvious. Was Christianity any more enlightened about scientific matters than those other world religions? Was the Bible an infallible work of science, or even a work of science at all?

"Darwin had to admit that the stories in the Bible are intended to be a guide for how to be a good person and for how people should get along with one another. They aren't really science at all.

"At this point, we all jumped into the discussion. We reminded Darwin that people may be reading the Bible too literally. Could any reasonable person believe that God went from one Galápagos island to another, distributing different species of plants and animals, including us bears, like some kind of cosmic gardener planting turnips? In addition, if God was truly responsible for every creation of a new species, was he also directly responsible for every natural calamity, including every animal and plant that dies from an accident or a disease? In other words, are we to believe that God systematically kills off some of the creatures he supposedly has created?

"If the Creator is truly an omnipotent bear, as some of us bears believe, don't you think he would have the power to produce species by means of natural laws rather than by performing one miracle after another? And is there any evidence that God actually likes to spend his time gardening and creating different animals and plants by miraculous intervention? After all, he would have had to perform an awful lot of miracles to account for all the different kinds of animals and plants around today. Why not just set evolution in motion and be done with it?

"I'm proud to say that our superior logic convinced Darwin that books like the Koran and Bible were never meant to be scientific guides for understanding how the world works. Science and religion are different ways of thinking, and if you want to know something about nature, science is the best tool to use. If you want to know something about being a good person and the many mistakes people make, then maybe a religious book would be helpful.

"Still, Darwin never became an atheist, as some people have supposed. He called himself an 'agnostic,' a word coined by Darwin's stalwart defender, Thomas Henry Huxley. As Huxley explained to us, this new word was derived from the Greek word for 'unknowable' and meant someone who simply does not know, one way or the other, whether or not there is a God. Darwin himself could argue the case for or against the existence of a deity. And he felt that believing in evolutionary theory didn't mean you couldn't also believe in God. He remained quite firm, however, in his conviction that God does not play an active role in the evolutionary process. That process, as he argued in his various books, is governed by natural laws, just as planetary motions are governed by Newton's law of universal gravitation rather than by miraculous guidance for each individual planet and its moons. And modern science overwhelmingly agrees with Darwin on this important issue."

Darwin Bear shifted in his seat as he grabbed another biscuit and sipped some more tea. Thus invigorated, he returned to his story. "I think I mentioned before that some of the bears changed their names sometime after we had reached England. You probably don't want to hear the reasons for each of these name changes, but two of them are worth mentioning because they are critical parts of the story of how strong the opposition initially was to Darwin's heretical ideas, as well as how his evolutionary theories finally won the day.

"In 1859, when the *Origin of Species* was published, Whewell Bear became quite upset when William Whewell—one of Darwin's Cambridge University teachers—expressed strong objections to Darwin's theory of evolution by natural selection, even refusing to allow a copy of the *Origin* to be kept in the Trinity College Library. Staunchly loyal to Darwin, this bear decided he would rather be called Wallace Bear, as a tribute to Alfred Russel Wallace.

"Wallace had independently hit on the idea of natural selection in 1858, while he was doing research in far-off Indonesia. In early 1858, he had sent a manuscript describing his own ideas about natural selection to Darwin, who was stunned to realize that his priority for this revolutionary theory was about to be forestalled. Fortunately, a reasonable compromise was reached, after the bears suggested that Darwin write to his good friends Joseph Hooker and

# Collected Essays
## by
### Harriet Martineau Bear

## Contents

**Chapter 1. How I Settled the Centuries-Old Conflict Between the Sea Lions and the Marine Iguanas**

**Chapter 2. On the Best Way to Avoid Being Engulfed by a Lava Flow: Run!**

**Chapter 3. A Proposal for Old-Age Social Security among the Galápagos Bears**

**Chapter 4. An Open Letter to the Feral Goats: Stop Eating Our Blueberries or Else!**

**Chapter 5. An Essay on Galápagos Bear Society, and Why You Might Want to Join It**

**Chapter 6. Letters to the Deaf: Advice to Bears and Tortoises about the Social Challenges of Hearing Loss**

**Chapter 7. A Reader's Guide to the *Rules and Regulations for Tortoise Racing***

**Chapter 8. Is Religion Good for You?: Definitely Yes, No, and Maybe**

**Chapter 9. What Thomas Malthus Got Right and Wrong in His *Essay on the Principle of Population* (1826 edition)**

**Chapter 10. Is William Paley's *Natural Theology* Really Natural? Ask Darwin.**

The first volume of Martineau Bear's *Collected Essays* was published in 1865, six years after the *Origin of Species*. This book is exceedingly rare, as only twenty copies were printed. These included a copy for Darwin and one for each of the bears. Martineau Bear presented two of the three remaining copies to the bears' good friends Joseph Hooker and Thomas Henry Huxley, with the third and last copy going to Harriet Martineau, to whom the book was dedicated.

Charles Lyell to ask their advice. Those two scientists arranged for the publication of excerpts from some of Darwin's unpublished writings, together with Wallace's manuscript, in the *Proceedings of the Linnean Society*. Wallace was such a nice fellow that he felt honored by this arrangement, and he always used to say that Darwin really deserved the lion's share of credit, because of the way in which the *Origin of Species* developed the ideas contained in their two brief Linnean Society papers, published in 1858. After his return to England, Wallace became a strong advocate of Darwin's theories, and he even wrote a book entitled *Darwinism* to honor Darwin's genius.

"The other name change occurred in 1860, as the bears all agreed that one of them ought to be called Huxley Bear, after Thomas Henry Huxley, who later came to be nicknamed 'Darwin's Bulldog' because of his spirited defense of Darwin's theories. This is a slightly longer story, but it is well worth telling.

"Huxley was an occasional visitor to Darwin's home in Downe, and we all enjoyed his scintillating wit and entertaining stories. Besides our weekly natural history walks with Darwin, we didn't get out much, so we kept on asking Huxley if he would show us around London—the zoo was a special favorite. Since Darwin and Huxley were both planning to attend the meeting of the British Association for the Advancement of Science in Oxford in June 1860, they suggested that five of us might come with them on the trip. The bears were particularly delighted that they would have a chance to see their friend Joseph Hooker, and they even brought some gooseberries to give to him. Huxley, for his part, clearly was raring for a fight, and had even written to Darwin to announce, 'I am sharpening my claws and beak in readiness' for the anticipated showdown over Darwin's controversial book.

"When Darwin got sick just before the Oxford meeting, Huxley agreed to look after the bears while he was there. So I, along with Hooker Bear, Wallace Bear, Henslow Bear, and Beagle Bear, all headed off to Oxford and stayed with Huxley in his rented rooms.

We listened to Huxley's defense of Darwin's ideas in a heated debate with Darwin's outspoken critic Richard Owen. It was all a lot of fun, but after the debate with Owen was over, Huxley was ready to pack his bags and go back to London.

"Just before our planned trip home, we happened to run into Robert Chambers on the street. Chambers had anonymously published, some years earlier, the notorious *Vestiges of the Natural History of Creation*, setting forth his own theory of evolution, amid fierce criticism. Even though Chambers's authorship was not made publc until 1884, Darwin had long suspected Chambers was the author and told us bears of his suspicion. When Chambers heard that Huxley was about to leave Oxford, Chambers begged him to stay for a showdown with Samuel Wilberforce, the Bishop of Oxford, who was expected to attack Darwin's theories the next day. We bears, of course, were all for staying and watching the fight, so we convinced Huxley to stay one more day to hear the pro-Darwinian talk by the American chemist and historian John William Draper, and the anti-Darwinian arguments of the bishop.

"Draper droned on for ninety minutes before a packed audience of more than seven hundred people and five bears. Hundreds more had been turned away from the overflowing lecture hall. Wearing broad-brimmed hats and dressed in our Sunday best, we bears were hardly noticed in the audience and were mistaken by some people for a family of dwarfs."

I almost interrupted here, to give Darwin Bear my own impression of him when I first saw him in the library. I thought he was a very short, very hairy child before I realized my mistake. And now here I was, one cold teapot and many platters of biscuits and sandwiches later, listening to his incredible story of his life with Darwin. "Ahem," I tried to interrupt, but the little bear was deep in his Oxford story and there was no stopping him.

"Finally Draper's boring talk was over." The bear yawned, apparently stupefied by the memory. Then he brightened. "Bishop Samuel

Wilberforce then stood up to denounce Darwin's theories. 'Is it credible,' he scoffed, 'that a turnip strives to become a man?' As an organizer of the British Association meetings, Huxley was seated on the platform with the other speakers. Knowing that Huxley was a strong supporter of Darwin's, Wilberforce turned directly to him at the end of his insulting speech and asked whether it was through Huxley's grandmother or grandfather that he personally claimed his descent from an ape!

"Huxley whispered to a man seated next to him, 'The Lord hath delivered him into my hands!' He rose, facing the bishop, and delivered a compelling defense of Darwin's theory while pointing out the many errors in Wilberforce's rant. He then answered the bishop's snide question with a zinger that has become famous in the annals of the history of science:

> If then the question is put to me whether I would rather have a miserable ape for a grandfather or a man highly endowed by nature and possessed of great means and influence and yet who employs those faculties for the mere purpose of introducing ridicule into a grave scientific discussion, I unhesitatingly affirm my preference for the ape.

"Huxley's retort to the Bishop elicited shouts, some jubilant cheers, and general pandemonium. One woman in the audience fainted and had to be carried out. Several other speakers, including Darwin's friends Joseph Hooker and John Lubbock, then added their own defenses of Darwin's theory. By the time the session was over, Darwin's defenders agreed they had won the day. It was the most remarkable scientific meeting that anyone could remember.

"Cheering wildly from the audience, Beagle Bear insisted on the spot that he was changing his name to Huxley Bear and that he would henceforth dedicate himself to becoming a great orator in

Thomas Henry Huxley responding to Bishop Samuel Wilberforce at the famous meeting of the British Association for the Advancement of Science (June 1860). *Left to right:* Huxley, Bishop Samuel Wilberforce, Benjamin Brodie, Joseph Hooker, John Stevens Henslow, and John William Draper.

the service of Darwinism, just like his namesake. Huxley, of course, was immensely flattered, and he later admitted to Darwin that some of his initial zeal in defending Darwinian theory stemmed from a secret desire to have a bear named after him. In his subsequent letters to Darwin and the bears, Huxley always appended the abbreviation 'FGB' to his signature—Friend of the Galápagos Bears. He considered this affiliation to be just as prestigious as the other scientific society honors that he always included after his name, on the title pages of his publications: FRS, Fellow of the Royal Society; FZS, Fellow of the Zoological Society; and FGS, Fellow of the Geological Society.

"Huxley Bear, in turn, kept up a lively correspondence with his namesake, sending him every tidbit he could find in the scientific literature that he thought might go over well in Huxley's publications or in his public speeches. Huxley Bear was extremely proud when one day Darwin said to him, 'You know, Huxley can be a real bulldog in supporting my theories—but you even outdo him: you're my grizzly bear!' "*

Darwin Bear sat back triumphantly, as if he were still at that famous meeting. Now, at last, I saw my chance to say something. But I was so dumbfounded by what I had just heard, and by how markedly it differed from the history I thought I knew, that I didn't know where to begin.

* For a debunking of the myth that Thomas Henry Huxley was widely known as "Darwin's Bulldog" during his own lifetime, see John van Wyhe, "Why There Was No 'Darwin's Bulldog': Thomas Henry Huxley's Famous Nickname," *The Linnean*, 35, no. 1 (April 2019): 26–30.

Karl Ernst von Bear receives the fateful package of haeckelberries.

## CHAPTER 7

# How Haeckelberries Almost Derailed a Revolution

—— or ——

# Why Psicko-analysis Failed as a New Science of the Mind

"You know, I've shown you how we helped Darwin," Darwin Bear said, resuming his discourse after a pregnant pause. "But I am embarrassed to say that at one point we almost derailed his scientific career and the whole Darwinian revolution with it."

"What do you mean?" That the bear was willing to admit mistakes made me a bit more willing to believe him. After all, nobody likes to talk about their own failures. Why make the boo-boos up if they didn't really happen?

The bear sighed. "Darwin's biographers have always wondered why he waited so long to publish his most important book, the *Origin of Species*—which did not appear until more than twenty years after his return from the *Beagle* voyage. Those biographers have provided all sorts of explanations for the delay, but no biographer has understood the principal reason.

"Yes, Darwin was a very careful scientist. He wanted to test as many of his arguments as possible, and he also felt it was necessary to check and recheck his facts and evidence before publication. But would all that really take twenty years?

"By the mid-1850s, Darwin had finished publishing his many findings from the *Beagle* voyage, including his brilliant work on barnacles, which earned him the Royal Society's Royal Medal in 1853.

He was finally ready to begin writing the *Origin of Species*—that is, until the unfortunate haeckelberry episode brought work on the book to a screeching halt."

"The *haeckelberry* episode?" I blurted out. There was no such kind of berry. "Don't you mean huckleberries?"

"No, I mean *haeckelberries*." Darwin Bear pronounced the word slowly and clearly, as if I were hard of hearing or a very small child. "Karl Ernst von Bear had been assigned the job of writing to various German scientists to gather information for Darwin's research. In these letters, he regularly included a request for information about local berries, particularly tasty ones. As Darwin was then busy with experiments testing whether seeds could survive immersion in salt water, the bears were eager to add berries to those experiments in order to prove that their Galápagos blueberries could have arrived by flotation rather than by miraculous creation.

"Karl had sent one of these letters to the famous German biologist Johannes Müller, who passed the letter on to one of his young students, Ernst Haeckel. Haeckel would later write directly to Darwin, a few years after the *Origin of Species* was published, to express his strong support for Darwin's ideas. But the letter to Müller gave Haeckel an idea about how to impress Darwin, whose book about the *Beagle* voyage he had read with great interest and admiration, and he scoured the countryside for any berries that might be interesting to Karl as well as to Darwin. This is why Karl received a package from Haeckel containing a '*gift—some very unusual berries worthy of further study.*' Karl began eating them, and he was soon spouting some extraordinary new ideas about evolution that he kept insisting Darwin must include in the *Origin of Species.*

"According to Karl's berry-inspired theory, every creature, during its own individual development, remembers and relives all the experiences of its ancestors. Of course, we were all rather skeptical that memories could be 'inherited' this way, but Karl kept insisting that he had discovered a new and profound scientific truth about the mind.

"Karl also thought that traumatic memories—especially those associated with having eaten bad-tasting berries, ones that had turned sour or moldy—could actually make us mentally ill. Painful memories, Karl believed, caused neuroses and psychoses, not to mention many of the little mistakes we all make every day. Karl called this universal source of sickness the 'Edible Complex.'

"We could be happy, according to Karl, only if each of these miserable experiences was fully remembered and thoroughly analyzed through a special process he called 'psicko-analysis.' Dreams were the key to recovering such bad memories. 'If you tell me your dreams,' he promised, 'I can cure you.' Personally, I felt no need of being healed. But he insisted that if we did not get rid of our ancestors' bad-berry traumas—if we kept those painful memories buried or tried to block them from our minds—we would never really be healthy and would continually show our underlying illness by making bad choices and stupid mistakes, even scientific mistakes.

"So every time a bear dropped a spoon or mispronounced a word, Karl would mutter, 'There, you see, it's another repressed memory at work.' And then he would say, while smugly puffing on one of the cigars he had begun smoking, 'You really do need psicko-analysis to retrieve these buried traumas and get rid of them all.'"

Darwin Bear shook his head sadly. I still felt far from understanding what haeckelberries had to do with this, but I listened patiently. "Soon Karl had each of the bears attending daily, hour-long sessions of psicko-analysis. Even Darwin had to go. We were supposed to report every detail about our dreams as some sort of window into the memories of our past lives.

"The problem was, we *didn't* remember any past lives. Karl, however, would prod us with a German accent he had begun using, 'Ahha, yust vhat I hed tought—*repression!*' Since we couldn't offer up any bad memories, Karl took that as evidence of how thoroughly we were quashing those memories. This of course meant we needed even more psicko-analysis. It was an endless cycle! To him, the fact

Karl Ernst von Bear tries to psicko-analyze the unconscious motives behind a large, messy inkblot, caused when Darwin's pen slipped from his hand.

that we couldn't remember was obvious proof that we were forgetting on purpose.

"The worst was poor Darwin. Karl insisted that Darwin's stomach-aches and bouts of dizziness, both of which had begun after his return from the *Beagle* voyage, were caused by his unresolved Edible Complex. As I have mentioned before, Darwin had a particular fondness for gooseberries, and he grew fifty-four different varieties of them in his garden. Karl was adamant that Darwin, or perhaps one of his ancestors, had repressed the painful memory of having eaten some bad gooseberries. Long-suffering Darwin was so desperate for help with his stomach problems that at first he believed Karl. Except the stomachaches continued and Darwin wasted a lot of time on the psicko-analytic sessions.

"Months went by, and work on Darwin's masterpiece, the *Origin of Species*, was falling further and further behind schedule. Yet after several hundred analytic sessions, neither Darwin nor any of us felt that we had recovered even a single real memory about eating bad berries. Nor were we any happier as a result of all this psicko-analyzing.

"Finally, enough was enough! Darwin argued that Karl's entire theory was unscientific because there was no way of ever disproving it. Karl saw every detail of our dreams as supporting his theory. If the details didn't fit, Karl would simply say something like, 'The unconscious mind has distorted the details of your dream into their opposites.' Or he would declare, 'Your dream means that you secretly want my theory of the Edible Complex to be wrong, which is unmistakable proof that it is right and that you are seriously ill. You clearly need a lot more analysis.'

"So we all got together to figure out what to do. Darwin asked us, 'Who knows anything about Ernst Haeckel's shipment of those haeckelberries that Karl has been eating lately? Does anyone know exactly *what* they are, *where* they came from, and whether they are really *safe* to consume? Maybe it's really Karl who has the Edible Complex, not us!' We then rummaged through Karl's letters from German scientists until we found the letter that had been sent with Ernst Haeckel's first shipment of the haeckelberries.

"Now here is where having a better understanding of German would really have helped. True, Karl Ernst von Bear had been named after a famous German embryologist, and he wrote constantly to German scientists, but he did not actually read or speak German fluently. He always used a dictionary to look up words he didn't understand. What Haeckel had written in German was that he was sending '*einige sehr ungewöhnliche Beeren, die es wert sind, weiter studiert zu werden, da sie ein Gift enthalten könnten.*' The key word *Gift* in this sentence means 'poison,' but Karl hadn't bothered to look up this word, thinking that *Gift* was the same in English as in

Darwin undergoes another useless session of psicko-analysis with Karl
Ernst von Bear.

German. Haeckel had actually been warning him about 'some
very unusual berries worthy of further study, which may contain a
poison.' What a terrible mistake! Karl had been eating the berries, thinking they were a tasty present when really they were quite
dangerous and addictive!"

Darwin Bear's eyes sparkled as he now explained: "With the discovery of Karl Ernst von Bear's terrible mistake, his strange behavior
finally began to make sense. The haeckelberries had been poisoning
his mind, causing him to have delusions about his own and other
people's past lives. So we threw away his remaining haeckelberries
and fed Karl a healing diet of tea and blueberries. Within a few days,
he was back to his old self again. Moreover, he stopped smoking
those smelly cigars, and even his German accent gradually faded
away. Karl himself could hardly believe all of the nonsense he had
been spouting about psicko-analysis and the Edible Complex."

"Karl was lucky that you figured this all out!" I said. "Could he
have died?" Obviously, by now these bears and their incredible story

were becoming real to me. For better or worse, I was beginning to believe almost every word Darwin Bear was saying. For a moment I thought, "Oh God, maybe I've eaten some haeckelberries myself!"

"If Karl had eaten more of the berries, he wouldn't have died," Darwin Bear replied, "but he would almost certainly have become permanently delusional. Luckily, he had eaten only a few each day. Ernst Haeckel did not fare as well. Although he suspected that the berries were poisonous and so warned Karl, he thought he could detoxify them. He ate his own test subjects and quickly became addicted, not caring whether the berries were safe or not. The haeckelberries affected his thinking so much that he concluded they were not toxic after all. As a result of Haeckel's continued consumption of haeckelberries, his thinking became permanently distorted. For example, he sometimes tended to prefer scientific speculation to any formal testing of his theories. Nor was he ever able to give up the deeply held conviction, initially induced by the haeckelberries, that human development from embryo to adulthood recapitulates the experiences of all of one's ancestors—which he announced to the world as his 'biogenetic law'—an idea that held considerable sway in late-nineteenth-century biological thought.

"Some years later, Haeckel shipped a large number of haeckel-berries to a young and ambitious Viennese medical student, who never suspected their toxic nature and also became addicted to them. This zealous fellow soon developed some strange theories of his own about the human mind—ideas that became wildly popular for many years in some circles and which he defended in book after book to the end of his life.

"As for the rest of us, we were able to get back to the *Origin of Species*. But now you know why the book took so long to finish."

"What about poor Darwin's stomachaches?" I asked, feeling a little queasy myself.

"The most likely cause is now thought to be a parasite that it appears he had picked up in South America. This parasite affects

After the unfortunate Haeckelberry episode, Darwin and the bears return to preparing the *Origin of Species* for publication. On the left, Caroline, Covington, and Henslow bears pick seeds from flowers and soak them in salt water—an experiment designed to show that plant seeds, unlike the eggs of amphibians, can survive long enough to be dispersed across the seas, even to very distant oceanic islands. Karl Ernst von Bear examines Darwin's collection of butterflies to see if the mouthparts are adapted to the specific flowers from which they typically obtain nectar. Emma, Henrietta, and Martineau bears, seated above the fireplace, are busy researching additional questions that need to be resolved before the *Origin* is finally published.

Darwin is using a microscope to confirm that frogs' eggs, collected from the terrarium, are killed by salt water. As Darwin later argued in the *Origin,* this finding explains the striking absence of frogs and other amphibians on oceanic islands. Seated on the desk next to Darwin, Darwin Bear records Darwin's observations, and Tennis Bear is asking if he can look through the microscope.

Gould, Huxley, and Lyell bears brush up on their reading, and Hooker and Humboldt Bears look on as Wallace Bear feeds the frogs in the terrarium. On the table, FitzRoy Bear is watering two orchids, which illustrate the structural coevolution of flowers and their pollinators—a subject to which Darwin devoted his next book after the *Origin*.

the heart and digestive system. Unfortunately, even today this disease is very difficult to cure if treatment is not begun right away, and no nineteenth-century doctor was ever able to help Darwin.

"Enough about stomach problems! All of this talk of berries has made me hungry for some really nice *blue*berries. Would you happen to have any for me and my friends?"

"Your friends?" I blurted out. "What friends?"

# PART FOUR

# The Gang's All Here

A small bear pops out of Darwin Bear's white bag. It's Wallace Bear.

## CHAPTER 8

# I Meet Darwin Bear's Friends

## — or —

# I Find Myself Shopping for Blueberries

Darwin Bear set down his teacup and began to open the white bag he'd brought with him. "My *bear* friends, of course!"

A furry little head popped out of the white bag. "This is Wallace Bear!" Darwin Bear exclaimed. "And these are Emma Bear and Fitz-Roy Bear," as two more heads poked out of the bag.

"You didn't think I would ever leave my friends behind in England, did you?" Four more bears clambered out of the bag. There were now seven bears crowded around Darwin Bear. They may have been patiently waiting before, but having been freed from the bag, they gulped down the last of the now cold tea and ate every remaining sandwich. But the bag was still wiggling, and eight more bears quickly followed, eager to lick up any crumbs left behind. The last bear was the tiniest of all. He was carrying an equally small tennis racket.

Emma Bear and FitzRoy Bear emerge.

The bag is finally empty!

"This must be Tennis Bear," I said. That was a name I certainly could remember! "And these must be all of the Galápagos bears."

"That's right." Darwin Bear nodded. "All sixteen of us. You may be wondering why we all came to America." (I was, in fact, wondering that very thing.) "We want to continue Darwin's work and we're here to do research for a textbook we're writing on the evolution of life on earth. After all, if we could help Darwin with so many of his own books, we thought we could write at least one good book ourselves. America, especially Harvard University, with its famous Museum of Comparative Zoology, has such wonderful libraries, we thought this would be an excellent place to start. That's why we were in the library today. But my story has taken so long that it's now past our lunchtime. Any chance you could offer us some *blueberries?*"

I didn't want to admit that I'd eaten the last bowlful myself that morning. Instead, I offered to run out and get some.

The bears all cheered. I felt a little strange leaving my home with sixteen bears in my living room, but the grocery store was a short walk away. How much trouble could they cause in just twenty minutes?

But twenty minutes can actually seem like a very long time when there are bears in your house. Would they start making long-distance telephone calls to England or the Galápagos Islands? Would they play hide and seek and upset all the potted plants? Would they go through the cupboards and eat everything they found?

Studious bears, busy rereading Darwin's collected works.

## CHAPTER 9

# The Bears Pursue Their Research

—— or ——

# Several Famous Harvard University Professors Are Quizzed about Bears and Blueberries

I raced back home with a bag full of blueberries, half expecting to find my place a shambles—only to find the sixteen bears seated quietly on the couch or in small groups on the floor, reading books from my shelves.

"What are you up to now?" I asked the bears.

"We're starting our research by rereading all of Darwin's works," Darwin Bear explained. "You have an excellent collection of his books."

"I never imagined that I would ever see such a studious group of bears!" I smiled.

"Good books are treasure troves," Darwin Bear said. "With a book like Darwin's *Journal of Researches* about the *Beagle* voyage, you can travel around the world and see many wondrous things without ever leaving your seat—except to get a few blueberry snacks now and then. Sometimes we each read a different book, and then we tell each other all about it. And sometimes one of us—usually Emma Bear—reads aloud to the others, just as Darwin's wife often read novels to Darwin in the evenings. Darwin encouraged us all to read, and life has been *so* much more exciting ever since."

"Well, how about a break for blueberry snacks now?" I said. "I'll make a fresh pot of tea for everyone." I never thought I'd be serving

I hold a tea party for the bears and offer them crackers with blueberry jam.

tea and blueberries to sixteen Galápagos bears, but here I was,
doing just that. This time, there was no long storytelling. The bears
were much too hungry to talk. I watched them stuff their bellies
with as much as they could possibly eat and drink (five pots of tea,
six extra-large containers of blueberries, blueberry jam on crackers,
and a few extra biscuits as well). When every bit had been eaten,
every drop swallowed, Darwin Bear announced that it was time for
them to get back to work. I thought that meant they would leave
and head back to the library, but they quickly went back to my own
shelves, pulling out more books to read.

I couldn't exactly kick them out. Besides, I was curious to see
whether they would reveal other nuggets about Darwin. So one after-
noon turned into two days, then three, and I eventually lost count.

The bears turned out to be very polite houseguests. They tidied up after themselves, didn't take long baths, or mess up the kitchen. They did eat a considerable amount of blueberries, but mostly they read.

Weeks and months went by, as the bears worked their way through my library of books by and about Darwin, books about evolution, and books about the Galápagos Islands. Usually the bears took notes on what they were reading, and sometimes they asked me to check out a few books from the Ernst Mayr Library at Harvard's Museum of Comparative Zoology. The bears were especially fond of books by Mayr, Edward O. Wilson, and Stephen Jay Gould, three of the most famous evolutionary biologists, all of whom had had offices in the Museum at one time or another.

Occasionally, posing as science reporters, the bears would telephone well-known biology professors, in particular Ed Wilson, and ask them difficult questions about various biological subjects, to see which one was the smartest. They asked questions such as "How did bears first evolve?," "Are whales related to bears?," "Why are blueberries blue?," "Why do bears love blueberries?," and "What is the difference between a blueberry and a gooseberry, and which makes a better pie?" Each of the scientists gave good answers, so the bears agreed that they were all equally smart—at least about bears and blueberries.

Although the blueberry questions were answered, the bears were surprised to discover that some of the foremost evolutionary biologists did not always agree on the best explanation for why things evolved the way they have. Take for instance the question of why blueberries are blue. According to Ernst Mayr, a world-famous ornithologist who wrote more than twenty books about evolution, birds tend to be attracted either to red- or blue-colored berries. Blueberries, in particular, are known to be especially rich in anthocyanins. These blue and red plant pigments have powerful antioxidant properties, protecting against age-related health risks and disease.

Research has shown that birds prefer berries whose colors suggest a high anthocyanin content. The plants benefit from producing these pigments because their brightly colored fruits, containing seeds, are then widely dispersed by the birds.

Mayr thought that the first blueberry to reach the Galápagos, millions of years ago, was almost certainly blue and had probably arrived as a seed in a bird's stomach. All subsequent Galápagos blueberries would have been blue because of what Mayr called the "founder effect." Had the first founder plant had greenish berries, for example, then all of its descendants would likely have had greenish berries too—assuming, of course, that evolution did not subsequently alter the color of the berries for some adaptive reason.

Mayr also explained that if the ancestral blueberry was not originally blue, it may have evolved its blue color in the Galápagos because of a "genetic revolution" in small populations. By this he meant that chance events play a much greater role in evolution within small populations, allowing the origin of novel traits and behaviors that might never evolve in larger groups. A good example is actually found in the Galápagos—namely, the vampire habits of the Sharp-beaked Ground Finch. This finch lives on two very small islands about 150 miles northwest of the main group. It has developed the unique habit of using its bill to open up wounds in the quills and skin of seabirds and drink blood from the wounds. Such behavior has not evolved in any of the other species of Darwin's finches living on other islands, including in other populations of the Sharp-beaked Ground Finch.

According to Niles Eldredge and Stephen Jay Gould, Galápagos blueberries would have been blue for a different reason. Eldredge and Gould were well known for the theory, proposed in 1972, that new species arise relatively quickly when they experience a new environment, and then generally remain unchanged for very long periods. In accordance with this theory, which Eldredge and Gould called punctuated equilibrium, Gould was convinced that an ancestral berry that was originally some other color had changed to blue soon after establishing itself in the harsh Galápagos environment. The blue color, Gould believed, had been preserved through a special botanical process he called "berry spectral inertia." When the bears informed Mayr about Gould's explanation, Mayr instantly blurted out: "Why, that's just another version of my own theory of rapid speciation following the peripheral isolation of small populations. That guy never gives me enough credit for my ideas!"

Edward O. Wilson, arguably the world's expert on the theory of island biogeography, told the bears he was skeptical about Eldredge and Gould's explanation for the color of Galápagos blueberries. Because these berries once grew on all of the islands—which, as noted, differ in size, height, local climate, and even species composition—Wilson thought punctuated equilibrium should also predict some color differences between the islands. Upon specifically questioning the bears on this point, Wilson found that this prediction was not supported by the facts.

Wilson offered the bears his own interesting theory about the blueness of Galápagos blueberries. He was famous for being able to predict the exact number of species living on oceanic islands, based on island size and distance from the nearest mainland. Smaller and more distant islands, Wilson's theory of island biogeography posited, tend to support fewer species and to receive fewer colonists as well. Wilson's ability to predict the number of species on individual oceanic islands sounded pretty scientific to the bears, and they were naturally eager for an equally exact answer to their own question about the color of blueberries.

To settle the question, Wilson even ran a few experiments for the bears. His experiments showed that berries with a blue color absorb low-energy red light, which lies at one end of the visible light spectrum, and they reflect back high-energy blue light from the other end of the spectrum. Strange as it may seem, light exerts a tiny amount of pressure on a surface (or at least that's what Albert

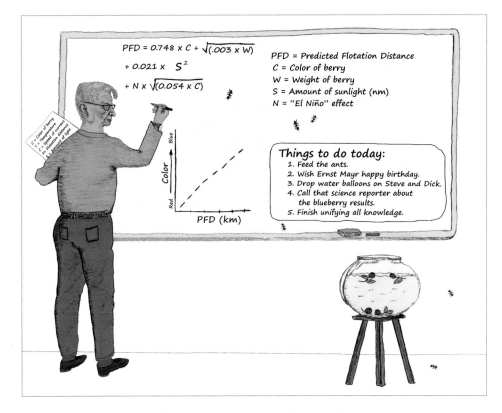

Having completed his experiments, Ed Wilson works on his formula to predict how far blueberries can float, depending on their color and other relevant factors.

Einstein and other very smart physicists have claimed), and the color of the light determines the amount of pressure. The bears weren't sure about all this, but because Wilson knew so much about biogeography they just took it for granted that he also knew a lot about the effects of different wavelengths of light.

In any event, Wilson hypothesized that berries having a blue color would be subject to less downward pressure from sunlight overhead and would therefore stay afloat much longer than berries that are red, yellow, or green. Thus they should be more likely than other kinds of berries to reach the Galápagos Islands by flotation. When

his experiments showed that the pressure exerted by sunlight is typically overcome by the forces of friction, Wilson realized that another factor must also be in play. He therefore postulated that floating berries having a dark blue color would be less noticeable to fish and seabirds and hence would have a lower chance of being eaten, compared to berries having lighter, more conspicuous colors. His tentative conclusions about expected flotation distance, he told the bears, were also supported by various mathematical formulas he had worked out. And to think that the Galápagos bears (who didn't know much mathematics) had always believed that blueberries were blue simply because blue is such a nice color!

When a similar mathematical formula predicted that ants should be either black, red, or brown, Wilson became even more convinced that he was right about the color of Galápagos blueberries. "There is a remarkable consilience in favor of my blueberry theory," he told the bears—by which he meant that the two sets of results reinforced each other, making them especially trustworthy. To drive home the importance of consilient lines of scientific evidence, Wilson sent the bears an Acme Company ant farm, along with a short note saying, "Dear Mr. Science Reporter, With the help of mathematics, I think I can now prove that ants are so widespread because their social organization is so consilient. If you observe this ant farm carefully, you will see what I mean." The bears were not entirely sure what this meant, although they guessed it had something to do with how successful ants are as a biological group because they practice cooperative social behavior. In other words, ants work toward the common good. The worker ants, who are all sisters, devote themselves to the welfare of the colony, which is headed by their mother, the queen.

The bears loved this magical-sounding word, "consilient," which seemed to represent the height of philosophical sophistication. The bears also noted the fact that the word was originally coined by the scientist and philosopher William Whewell, after whom one of the bears had been named. (Whewell also coined the word "scientist.")

Darwin Bear and Tennis Bear, feeding blueberries to their pet ants.

After getting Wilson's note, the bears wrote a song that included every word they could think of that rhymed with "consilient." Here are the verses to their song, which they set to the Scarecrow's performance of "If I Only Had a Brain," in *The Wizard of Oz*.

### The Consilience Song

1.  Ants are so resilient,
    Because they're so consilient,
    Or so says Dr. Ed.

    If the world paid more attention
    To consilient comprehension,
    Then we'd never be misled!

2.  The goal of wise Consilience
    Is definitely brilliance,
    On this we'd all agree.

    For Darwin it was finches,
    Giant tortoises, and fishes,
    Topped by pigeon pedigrees.


3.  When bears achieve Consilience,
    They think with greater diligence,
    In all they try to do.

    With consilient examples,
    The truth is never trampled,
    And blueberries must be blue!


4.  Consilient induction
    Needs no introduction,
    Thanks to Dr. Ed.

    But Whewell deserves some credit
    Since he had also said it,
    Making him the first instead.


5.  Consilience is magic
    When experiments are classic,
    And discoveries are new.

    But it's crucial to remember,
    That to be a true contender,
    There is always more to do!

After the bears couldn't think of any more words rhyming either with "consilient" or "consilience," they turned their attention to drawing up an evolutionary tree of bears. And here you see it:

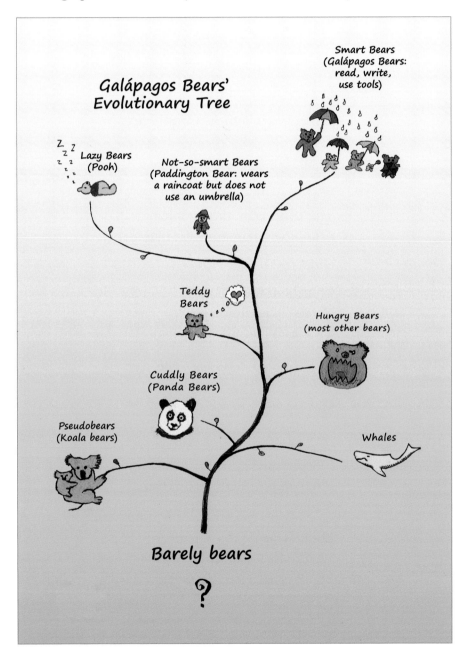

# PART FIVE

# The Return Home

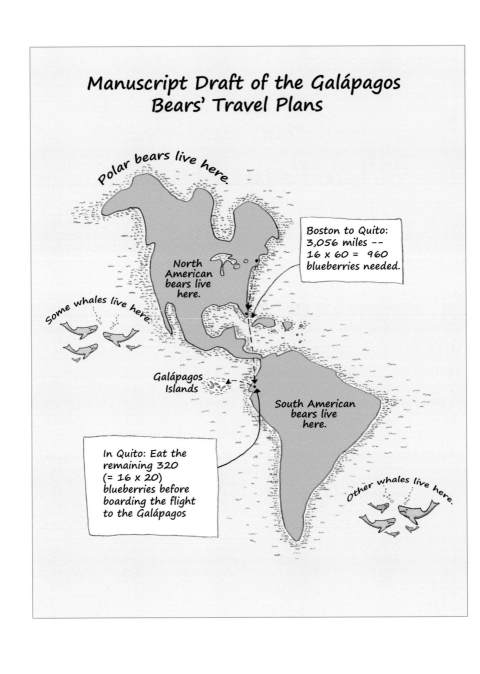

# I Am Invited to Join a Scientific Expedition
### —— or ——
# The Bears Overcome Serious Obstacles and Finally Return Home

O ne day, many months into the bears' rigorous program of read-ing and research, I arrived home to find them all gathered in a circle and having a heated discussion.

"We are planning an expedition," Darwin Bear cried out, as he greeted me. "Would you like to come along?"

"An expedition to where?" I asked, although I had a sneaking suspicion.

"Back to the Galápagos Islands!" Darwin Bear announced. "We have been reading about all those unwanted plant species that have been introduced to the islands by humans, like blackberries. They're now overrunning the highlands on Santa Cruz and other islands and choking out much of the native vegetation. We've come up with a really good solution to the blackberry invasion—one that only we bears could accomplish. And while we're eliminating blackberries, we would also like to retrace Darwin's footsteps in our archipelago so we can write a book about how we assisted him during his visit. In short, we want to solve the blackberry problem and correct all the silly mistakes made by Darwin scholars who don't know the crucial part that we bears played in his life."

"I'd love to join you!" I said. "But do you have all the necessary research permits?"

"What permits?" Darwin Bear asked. "Do you mean we need official permission to go back to our own home?"

I explained that visiting the Galápagos Islands is strictly regulated by the Galápagos National Park Directorate, which, together with the Charles Darwin Research Station, is trying to protect the islands against further ecological destruction and, through special breeding programs, to save various endangered species, such as tortoises, land iguanas, and some of the finches. "Before their final journey to the archipelago," I added, "all visitors are subject to careful inspection to ensure that prohibited organisms are not brought to the islands. Visiting scientists must submit their research proposals well in advance of being allowed to go into the field."

I offered to write a proposal for the bears, listing them as "research assistants" who had all previously worked with "an eminent British evolutionary biologist." Of course I meant Darwin, although I didn't mention him by name for fear that the Galápagos National Park Directorate would think our application was a hoax.

Several months later, the National Park Directorate granted us permission to come to the islands to carry out our research proposal. I made arrangements for myself and all of the bears to fly to Quito, Ecuador, and then on to the Galápagos. The bears packed a plentiful supply of blueberries, as well as some of Darwin's favorite variety of gooseberries, which they called "Darwinberries." They also packed a small supply of Galápagos blueberry seeds, which they had used to grow blueberries in Darwin's garden in England, and which they hoped they might now reintroduce to the Galápagos.

I warned the bears that, unfortunately, we couldn't bring any fruits or seeds with us. They would have to eat everything before we left Quito, or their seeds and berries would be confiscated at the airport by the Quarantine Inspection System for the Galápagos. The bears were quite upset by this news, so I suggested that we try to obtain official permission for a visiting scientist from America to hand carry some of their seeds to the Galápagos at a later date.

The bears catch their first glimpse of land in the Galápagos, after so many years away.

A colleague who regularly does research in the Galápagos kindly offered to bring the bears' seeds during her next trip, if permission was granted.

As we began to see some of the islands from the air, the bears started jumping up and down in their seats. They peered out the windows, calling to each other, "Oh, look—there's Huxley Bear's island, and there's Hooker Bear's Island. And look, on the other side of the plane, there's Darwin Bear's island." This was a real homecoming!

We stepped off the airplane into the intense midday equatorial heat. All around us was a flat, lava-strewn landscape, studded with giant *Opuntia* tree cacti bristling with spines, leafless Palo Santo trees, and thorny *Parkinsonia* bushes. I was reminded of the description of the desolate Galápagos landscape by Captain Robert FitzRoy, who wrote of "a shore fit for Pandemonium." Darwin himself, upon his first arrival, remarked that "the country was compared to what we might imagine the cultivated parts of the Infernal regions to be"—a fancy way of saying that the archipelago looked like the devil's home. In spite of those bleak descriptions, there is a stark beauty to such volcanic vistas that grows with each visit.*

We had landed on Baltra, a low, flat island, and once the home of Karl Ernst von Bear. From the tarmac, Karl pointed out where some of his favorite napping spots used to be, where his closest land iguana friends lived, and where he used to find the largest blue-berries. Karl was surprised to see a number of concrete foundations for former buildings, the remains of a large U.S. military presence on this island during World War II, when Baltra was leased from the Ecuadorian government as an airbase to protect the Panama Canal. After we walked into the airport terminal, I presented our official letter of invitation to one of the inspectors, whose job it was to make sure no nonnative plants or animals are brought into the archipel-ago. The inspector read the letter and then stared long and hard at my sixteen assistants, who were lined up smartly behind me.

"Aren't those *bears?*" the inspector demanded.

"Why, yes," I replied. "But they were born here, and they won't be any trouble at all."

"What do you mean, *bears were born here?* No bears have ever been born here! Or lived here! Or died here!" the inspector barked back. "And they certainly are *not* going to be allowed on the Galápagos

---

* FitzRoy's description of the Galápagos, along with his own account of the *Beagle* voyage, was later published as *Narrative of the Surveying Voyages of His Majesty's Ships Adventure and Beagle, between the Years 1826 and 1836* (London: Henry Colburn, 1839). Darwin's comparison of the Galápagos to the "Infernal regions" was recorded in his voyage *Diary* on the first day of his five-week visit (September 15, 1835). See *Charles Darwin's Diary of the Voyage of H.M.S. "Beagle,"* edited by Nora Barlow (Cambridge: Cambridge University Press, 1933).

Karl Ernst von Bear points out some of the interesting sights on Baltra. This small island consists of a flat region of lava rubble formed by sub-terranean volcanic forces that caused faulting and an upward thrusting of the sea floor.

Islands as visitors. Their presence is already a serious violation of our National Park rules, since they are *not* native animals. I'm sorry, but you will all have to come with me."

The inspector led us away, accompanied by several other stone-faced men who were also wearing Galápagos National Park uniforms.

As we walked through the airport under their guard, tourists in the airport began whispering to one another and pointing at us.

I was separated from the bears and taken to the inspector's office, where in vain I tried to convince him that he should allow me and my sixteen assistants to do our already approved research. I waved papers, certificates, passports, all manner of official paperdom at him.

He stood firm, insisting that he had to follow National Park guidelines as well as the quarantine rules of the Galápagos Biosecurity Agency. The National Park director had been called, he said, and the director would deal with us. But there was no way he'd let us in, the inspector warned. "Simply unheard of," he mumbled under his breath. "What's next—lions, tigers, alligators, and elephants?"

I asked the inspector if I could be with my friends. He nodded, and several guards led me to a large holding cell, which had metal bars along the front wall and on the only window at the rear. It was a sad sight to see, but the bears themselves were doing their best to make light of the situation, singing songs and telling one another— and the occasional visiting finch or two—just how nice it was to be back in the Galápagos, even in a jail cell.

That was true for an hour or so, but many hours went by. Then several days. No one had told us that the National Park director was doing fieldwork on another island, where the park was repatriating young tortoises raised at its tortoise breeding center, in an effort to save some of the endangered populations from extinction. So more days went by as we anxiously waited to learn the fate of the bears.

"What is going to happen to us?" moaned Darwin Bear. "Don't they know this is our home?" After the third day, the bears began plotting an escape. The guards had overlooked the fact that Tennis Bear, who looked more like a small toy than a real bear, could easily slip through the bars of the cell and steal the keys to the cell door. Once they all made their getaway, they would call the sea turtles to take them to one of the uninhabited islands.

Darwin Bear and his friends in their holding cell, discussing possible
ways to escape.

I pointed out that the bears would then be fugitives and that
National Park rangers would likely track them down with trained
hunting dogs. They might even shoot them. Was this a risk worth taking? I still hoped to convince the National Park director to let us all
go, once we had the opportunity to meet with him and plead
our case.

The bears didn't at all like the idea of being tracked or shot. So
they began to work on a different plan. Plan B was to hypnotize the
guards and then plant a posthypnotic suggestion in their minds that
"bears have always lived in the Galápagos," "bears belong here," and
"bears are your friends." Karl Ernst von Bear, it turns out, was quite

good at hypnosis—the one useful thing he had learned during his studies of psicko-analytic therapy.

I agreed that this plan could work, but I still wanted one last chance to talk to the National Park director. On the fifth day, following his return from the field, the director finally came to decide our fate. To our considerable relief, he was a kind man who also had an excellent reputation for devotion to Galápagos conservation. He had grown up on Floreana Island, like FitzRoy Bear, and he held degrees in biology and environmental management.

"What's this I hear about *bears* trying to sneak into our National Park?" the director asked me. "We already have enough trouble with the feral pigs, goats, donkeys, cats, dogs, and rats that shouldn't be here. We certainly can't have bears!"

He peered at me suspiciously. "Are you a friend of that Ecuadorian conservationist's brother, the one who wanted to fill the Galápagos with lions and tigers? He claimed they would be good for the islands for two reasons. First, they would kill all the feral pigs, donkeys, and goats. Second, they would generate much-needed tourism by attracting big game hunters, who would sign up in droves to go on Galápagos safaris to hunt these big cats. You aren't so misguided as to be proposing bear safaris, are you?"

"Absolutely not! These aren't bears to be hunted!" I protested. "They are here to do important research, and they have every right to be here, since they were born here. And, yes, lived here. These are the last remaining *Galápagos* bears, native to these islands. Their ancestors lived here for millions of years before humans arrived. Not only are they very conservation-minded, but these particular bears actually met Charles Darwin during his famous visit here in 1835. Darwin personally saved them from certain extinction by taking them back with him to England, along with the last of the indigenous blueberries on which they survived. These bears therefore represent the first successful conservation program in the history of the islands! Given their intimate knowledge of the islands, as well as their many years of study with Darwin, they could be of great help to you in your own conservation efforts."

Several of the bears began singing the first verse of their song: "We Bears love Charlie D., It comes so naturally, He saved us Bears. . . ."

"Not now," I whispered, signaling them to cut the song short. Turning back to the National Park director, I said, "One of these bears lived on Floreana, the island where you grew up," and I pointed to FitzRoy Bear, who nodded his head, tipped his naval cap, and saluted smartly.

I had forgotten how long it had taken me to believe Darwin Bear when he first told me his story. Would the director think I was crazy? How could I prove any of this to him? What scientific evidence did I have, after all? Had we traveled all this way only to be thrown off the island like criminals? I had never so much as gotten a parking ticket, and now this director was sizing me up as if I were someone guilty of much worse.

I looked hopefully at Darwin Bear. Could he convince the director the way he had won me over?

But Darwin Bear didn't say anything (which was not like him at all). None of the bears did. They were as worried as I was.

The park director stared intently at each bear. When he finally came to Tennis Bear, he bent down to take a closer look. "Little bear, may I please examine your tennis racket?" he asked. Tennis Bear grinned and eagerly handed over the racket, perhaps hoping they might play a game together. The director took a magnifying glass from his shirt pocket and inspected the racket closely, as if it were a fascinating scientific specimen and not a piece of sports equipment.

After several long minutes, as I was imagining our hasty return to Quito, the director looked up, his face lit with excitement. "I have seen tiny rackets like this one before—in fact, I've seen quite a few of them. A paleontologist looking for fossil bones—his name was Steadman, I believe—recently found some of them on Darwin, the

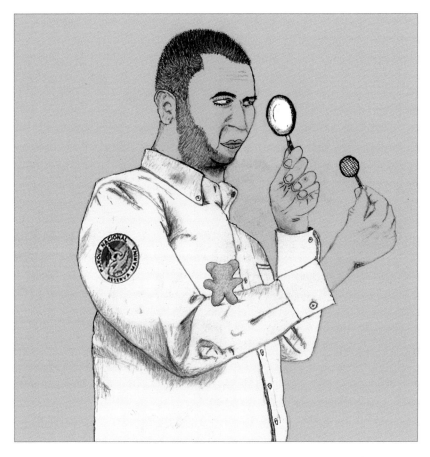

The National Park director carefully examines Tennis Bear's racket.

northernmost island in the archipelago. Almost all of these rackets have curious purple stains on their frames and strings. The park has just completed a DNA analysis of these stains, and the results indicate that they were caused by some kind of blueberry juice, though the DNA profile appears different from that of any known blueberry plant in the world. It's apparently a new species to science—one that originated in the Galápagos and has since become extinct."

"Just as we said!" Darwin Bear crossed his arms smugly.

The director took another look at Tennis Bear's racket. "What is even more curious about this recent paleontological discovery is that some of these tiny tennis rackets were found covered over by lava flows that appear to be tens of thousands of years old and clearly predate any human habitation of the islands. This means the rackets could not have belonged to humans, who first visited the islands in the sixteenth century, but must instead have been left here by some other creature. It is known, of course, that the Woodpecker Finch uses small twigs to help it pry insects from crevices in tree trunks, making it one of the very few tool-using birds in the world. I concluded that these tiny rackets proved there must once have been another tool-using bird here in the Galápagos. I never dreamed that the rackets might once have been used by small *bears*."

Darwin Bear now finally launched into the whole saga of how every island in the Galápagos was formerly inhabited by its own kind of bear and how, on Darwin Island, all the tennis bears had adapted to their sterile surroundings and meager berry supply by becoming almost as small as the blueberries. He was about to begin an account of how the bears got their names, the problems with psicko-analysis, and Darwin's stomach problems when I nudged him with my foot. He got the hint and quickly wrapped things up by describing Darwin's rescue of the last of the Galápagos bears.

The director stroked his chin thoughtfully. "Strange as your story sounds, I find myself almost convinced. But if you are really the 'Galápagos bears' you claim to be, and if FitzRoy Bear really did once live on Floreana, then he ought to be able to answer this question: What kind of giant land tortoise lived on Floreana before becoming extinct soon after Darwin's visit there? More specifically, did this particular tortoise have a dome-shaped form or a saddleback form?"

"Why, a saddleback, of course!" FitzRoy Bear blurted out. "I know lots more than that! In fact, I know almost everything there is to know about Floreana. For example, the Galápagos petrels like to nest on the tallest hill, called Cerro Pajas. And near the Governor's Dripstone in the highlands, where there is fresh water, there are

A Woodpecker Finch, using a tool to probe for insects. This remarkable species is thought by the bears to have evolved after a finch attended a tennis class taught by one of the tennis bears. (After a photograph by Roger Perry)

pirates' caves where I used to live with some of my pirate friends. And there is a big cave about a quarter of a mile east of Post Office Bay that contains the bones of various animals—mostly tortoises—that accidentally fell into it over the millennia. Ask me anything at all about my island. I bet I know a few things about Floreana Island that even you do not know. For example, I know that the distinct saddleback form of tortoise that once lived on this island, and that has long been considered extinct, has actually survived and currently lives on another of the Galápagos Islands. Sometime during the early nineteenth century, a whaling ship took on such a large catch of whales that it had to make room by offloading several Floreana tortoises, which had been brought on board for food. These poor tortoises were thrown into the ocean. Some of them managed to swim to safety at nearby Volcán Wolf, the northernmost volcano of Isabela Island. We bears learned about this miraculous tale of survival a few years later, after one of our Tour de Galápagos tortoise races was held on Isabela. Because tortoises live so long, it is likely that some of the tortoises living near this volcano today will have a very high proportion of their genes inherited from the original Floreana species."

"Now I *am* convinced!" The director smiled broadly. "The National Park just received a news bulletin from a distinguished team of scientists confirming exactly what you have said about the possible survival of the Floreana tortoise, and these exciting results have not even been published yet. I have to admit I'd heard rumors about bears once living in the Galápagos, but I always believed this was just a fanciful legend, like those about dragons, mermaids, and the Abominable Snowman. But Tennis Bear's racket looks authentic, and the stains found on the recently discovered Darwin Island rackets are definitely from blueberry juice. Finally, FitzRoy Bear's comments about my island are entirely correct."

The director tucked his magnifying glass back into his pocket, handed the tiny racket back to Tennis Bear with a deferential little bow, and asked, "So, what has brought all of you bears back to the Galápagos Islands?"

The Galápagos bears eating their way through the blackberries in the Santa Cruz highlands.

# CHAPTER 11

# The Bears Undertake
# an Ingenious Conservation Project

——— or ———

# Darwinberry Jam Becomes a
# Big Hit in the Galápagos

The room—so tense before—seemed suddenly filled with fresh air. The director ordered the immediate release of all the bears from their jail cell. The inspector scowled, but he had to obey.

"Please do tell me why you've returned after so many years," the director repeated, once the bears had been fed and allowed a chance to brush their teeth after their long days behind bars.

Darwin Bear, of course, answered. "We are all part of a research and conservation team. We have a plan to get rid of all the blackberries in the archipelago, and we also have some ideas that might assist you in your other conservation efforts." He left out the bit about the books they wanted to write. I suspected he would give the director an earful about that later.

"How can you possibly get rid of blackberries in the Galápagos?" the director asked. "The National Park Directorate has tried to do so for years, and the most we've managed to do is contain their spread. Blackberries are one of the most invasive plant species on the islands, growing out of control throughout the highlands on Santa Cruz Island as well as on some of the other islands, where they have created great thorny tangles. Some farmers have become so frustrated by this blackberry epidemic that they've abandoned their fields and moved to the lowlands to work in the tourist industry."

"It's as simple as pie!" Darwin Bear smacked his lips. "We plan to pick and eat all the blackberries as they ripen. Then there will be no seeds to create new blackberry plants. And because blackberry plants also reproduce by means of runners, we will clip off all of the runners as we harvest the berries. Without ripe blackberries to provide seeds for the next generation, the mother plants won't be able to spread, and the existing plants will eventually die out completely, just as our beloved blueberries did. Blackberries will be completely eliminated from the ecosystem! Thanks to our friend Darwin, we bears have acquired a voracious appetite for blackberries—although we *do* prefer blueberries—so we should be able to eat our way through them pretty quickly. At least, we'd like to try. We call our plan 'Extinction of the Yummiest.'"

During the next hour or so, the bears and I reviewed our field-work proposal with the park director. And that was that! We were granted entry into the islands and permission to begin our conservation efforts. The director also offered to help us in any way the park service could. As an apology for their time behind bars, he invited the bears to the park headquarters on Santa Cruz Island, where he presented each bear with an official identification badge saying "Warden, Galápagos National Park Directorate—Special Blackberry Eradication Squad." He even had a tiny badge made especially for Tennis Bear. It was almost as big as Tennis Bear's head, but he wore it proudly.

The first thing the bears did before venturing into the field was to learn the rules and regulations created by the Galápagos National Park Directorate to protect the islands and their native species. The bears were surprised to learn that tortoise races were no longer permitted, as tourists are required to remain at least two meters away from animals. Accordingly, they revised their *Rules and Regulations for Tortoise Racing in the Galápagos* to reflect the new regulations (Appendix 2, pp. 135–36). The major change they made in this document was for each bear to be represented by a prospective champion tortoise, which the bears would select and train, with all races now to be run by the tortoises alone, no riders involved. The tortoises were happy to hear about this rule change and even offered to give up one of their official weekly days of rest, on the first Tuesday of each month, to get themselves in better shape for the races.

Over the next few months, the bears guided me through each of their original island homes, showing me exactly where Darwin had collected specimens on each of the four islands he visited so long ago. Then they returned to Santa Cruz Island to begin their program of eradicating blackberries from the highlands of this badly infested island. As the months passed, the bears were so successful at gathering blackberries that they had far more than they could eat (and they could eat a lot of berries, believe me). So they began making jam from the leftover blackberries, using a special formula[*] based on a recipe from Emma Darwin's recipe book. Under the label Darwinberry Jam, the bears sold their jam in the town of Puerto Ayora, where it quickly became a favorite with tourists, especially those on the large tour ships.

---

[*] See Appendix 4 (pp. 145–46) for the recipe for Darwinberry Jam.

Covington Bear, Hooker Bear, and Emma Bear were put in charge of sales, inventory, and marketing, respectively. As the Darwinberry Jam business boomed, the bears donated the profits to the Charles Darwin Research Station. The station, in turn, agreed to provide them with room and board. While the bears were engaged in eliminating blackberries on Santa Cruz, they also worked hard at uprooting various other invasive plant species, such as the quinine trees that had overrun much of the highlands, endangering the nesting sites of the Galápagos petrel by choking off the native *Miconia* shrubs, under which the petrels liked to nest.

The bears were so successful that they eventually ran out of surplus blackberries for Darwinberry Jam. Luckily for them, the National Park had finally given the bears permission to import their native blueberry seeds, which were delivered by the visiting scientist to whom I had entrusted them. By the time the bears' supplies of blackberries had dwindled, their blueberries were flourishing in an enclosed highland site even as the park was considering the matter of whether blueberries might ever be reintroduced to the islands. The bears substituted their native berries for blackberries and called this jam Darwinberry Jam Deluxe. It proved even more popular than the previous product.

When in Puerto Ayora, some of the bears occasionally volunteered to work at the research station's library, where they helped catalogue the ever-expanding Galápagos research literature published by resident and visiting scientists. There they created a special library shelf labeled "Research on, or by, Galápagos Bears," on which they placed a copy of the *Origin of Species*, a book about whales, and Martineau Bear's personal copy of her *Collected Essays*. On the same shelf, they also stored a boxed record of their own considerable research and conservation efforts. At night, the bears generally slept cuddled up together in the rafters of the library. There is no more comfortable place to sleep, they thought, than among books.

Making Darwinberry Jam Deluxe.

Tourists may have stared at them, but aside from cultivating a few close friends at the Darwin Station and the National Park Directorate, the bears generally kept to themselves. Although they were careful to avoid most local residents and tourists, they sometimes made an exception when they met groups of children. Then they would often pop out from behind cactus trees or lava boulders to say hello. The bears especially loved to launch into stories about their own adventures; the kids loved to hear them and had no problems believing the bears right away. If only I had been as smart as those children! When I think back to that first day when I saw Darwin Bear in the Harvard library, I'm embarrassed to admit that I thought he was a consummate liar.

After about six months, I had to return home to write up the results stemming from my time in the archipelago. The bears, it was decided, would stay where they had always belonged—in the Galápagos. They were doing good work, and they were back with their old friends—the tortoises and finches.

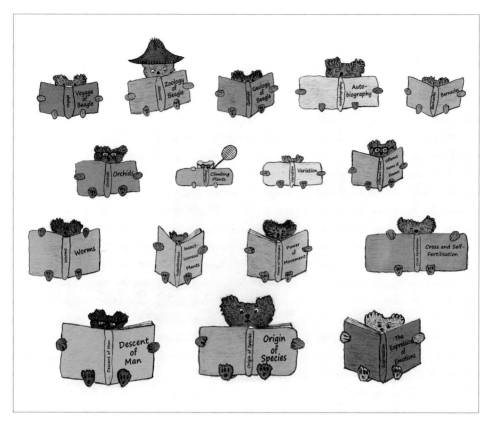

The Galápagos bears with the book by Darwin to which each made the greatest contribution. *Top row (left to right):* Covington, FitzRoy, Lyell, Emma, and Gould bears. *Second row:* Karl Ernst von Bear, Tennis, Henrietta, and Hooker bears. *Third row:* Humboldt, Caroline, Henslow, and Huxley bears. *Fourth row*: Wallace, Darwin, and Martineau bears.

It wasn't easy to say good-bye. I had become quite fond of all the bears. Of course, I promised them I wouldn't leave out their crucial part in the story of Darwin's scientific discoveries. And I put together a nice good-bye party with the help of friends from the Darwin Station and the National Park. As might be expected, the menu consisted of blueberries, blueberry muffins, blueberry pie, blueberry juice, and plenty of Darwinberry Jam Deluxe. On that last morning, before I left for the airport, I hugged each bear good-bye, Darwin Bear longest of all.

"Go home and write," he said. "The world needs to hear our story!"

And that's exactly what I've been doing—writing down their story, as you've been reading it here.

Every year more than two hundred thousand visitors come to the starkly beautiful Galápagos Islands to see—among other remarkable creatures—the giant land tortoises, land and marine iguanas, and Darwin's finches. But as we now know, these islands are also a very special place where sixteen different kinds of bears may once again be glimpsed by those who know just where to look for them.

### *The End*

# Appendices

# All the Galápagos Bears, Their Habits, Talents, and Islands*

### Caroline Herschel Bear · *Tower/Genovesa*

Especially knowledgeable about the stars and constellations, and an excellent observer and teacher. Named after Caroline Herschel, an accomplished astronomer who catalogued nearly 2,500 nebulae and discovered many comets. She was also the aunt of famed natural philosopher John Herschel, whose own astronomical researches she assisted and whom Darwin greatly admired.

### Covington Bear · *Chatham/San Cristóbal*

A very industrious bear, likes to collect all sorts of things. Named after Syms Covington, a fiddler and cabin boy on the *Beagle* who became Darwin's assistant and later his personal servant.

---

* Because Darwin's scientific friends and role models tended to be male, he suggested more male names for the bears than female names. This didn't bother the bears one bit, as they just wanted to be named after someone important in Darwin's life. So four of the eight girl bears (Gould, Henslow, Humboldt, and Lyell) adopted male names.

### Darwin Bear · *James/Santiago*

Observant and theoretically inclined; also, particularly knowledgeable about the Rules of Tortoise Racing (Appendix 2). A favorite among the other Galápagos bears, and their acknowledged leader. Named after Charles Darwin.

### Emma Darwin Bear · *Jervis/Rábida*

Gifted at foreign languages, musically talented, and an excellent nurse. Devised the recipe for Darwinberry Jam. Named after Darwin's first cousin, who later became Darwin's devoted wife.

### Finch-Bear · *Tower?/Genovesa?*

Known only from folklore, the Finch-Bear is thought to have evolved on Tower Island after a bear fell madly in love with a finch and they had baby bird-bears. All too successful at finding blueberries, the hybrid finch-bears became extinct after consuming all of the blueberries on their island. Tower/Genovesa Island was later colonized by the ancestors of Caroline Herschel Bear sometime after the extinction of Finch-Bear.

### FitzRoy Bear · *Charles/Floreana*

An accomplished if sometimes irritable sailor. Also a keen student of sea creatures, and particularly fond of speculating about the relationship between whales and bears (Appendix 3). Named after Robert FitzRoy, captain of H.M.S. *Beagle*.

## Gould Bear · *Wenman/Wolf*

Artistic, and very knowledgeable about birds in the Galápagos and other places. Likes to give scientific-sounding names to things. Named after ornithologist John Gould, who helped Darwin understand the full implications of his *Beagle* collections of birds.

## Henrietta Darwin Bear · *Abington/Pinta*

Very good at grammar and punctuation, the best proofreader among the bears. In addition to assisting Darwin with his many experiments, she helped edit *The Descent of Man* (1871). Named after Darwin's daughter Henrietta. Formerly known as Elizabeth Gould Bear, after the talented artist and wife of ornithologist John Gould.

## Henslow Bear · *Narborough/Fernandina*

A botanist and superb teacher. Named after Darwin's beloved mentor, botanist John Stevens Henslow, who arranged for Darwin's appointment as naturalist on H.M.S. *Beagle.*

## Hooker Bear · *Indefatigable/Santa Cruz*

Somewhat shy and reflective, and, like Henslow Bear, very knowledgeable about plants. Named after Darwin's close friend Joseph Dalton Hooker, director of Kew Gardens in London and, like Darwin, a lover of gooseberries. Previously known as Wickham Bear (after the *Beagle*'s Lieutenant

John Wickham), he may have been bribed by Hooker to change his name with the gift of some extra bananas ("Kew gooseberries") grown at Kew Gardens.

**Humboldt Bear** · *Hood/Española*

The most traveled of all the bears, knows all of the Galápagos Islands well. Named after the famous explorer Alexander von Humboldt, whose seven-volume *Personal Narrative* (1819–1829) of his travels in South America between 1799 and 1804, along with John Herschel's *Preliminary Discourse on the Study of Natural Philosophy* (1831), stirred in Darwin "a burning zeal to add even the most humble contribution to the noble structure of Natural Science."

**Huxley Bear** · *Barrington/Santa Fé*

Dashing, quick-witted, and charming; the best public speaker among the bears. Named after Darwin's friend and staunch defender Thomas Henry Huxley. (Before 1860, known as H.M.S. Beagle Bear.)

**Karl Ernst von Bear** · *South Seymour/Baltra*

Pensive and philosophically inclined. Dabbles in hypnosis and also likes to try new foods. Named after the German embryologist and explorer Karl Ernst von Baer.

**Lyell Bear** · *Albemarle/Isabela*
Sociable and talkative; also the keenest student
of rocks among the bears. Named after Darwin's
friend the eminent geologist Charles Lyell.

**Martineau Bear** · *Bindloe/Marchena*

Knows a remarkably wide range of controversial subjects.
Well read, open-minded, and the author of numerous pamphlets on social and religious issues, which
she regularly circulates among the bears. Named
after Harriet Martineau, a writer, pioneering femi-
nist, and social scientist greatly admired by Darwin.
Martineau was also a close friend of Darwin's elder
brother, Erasmus.

**Tennis Bear** · *Culpepper/Darwin*
The smallest of the Galápagos bears. Has quick
reflexes and is passionate about tennis. Lively,
friendly, and outgoing. Named for the racket he
always carries.

**Wallace Bear** · *Duncan/Pinzón*

Like Darwin Bear, adept at coming up with new and
fruitful theories. Also known to be adventurous,
hardworking, modest, and well liked. Named after
Alfred Russel Wallace, who, independent of Darwin,
formulated his own theory of natural selection in 1858.
Before 1859, known as Whewell Bear, a name he discarded in pro-
test after William Whewell opposed Darwin's evolutionary theories.

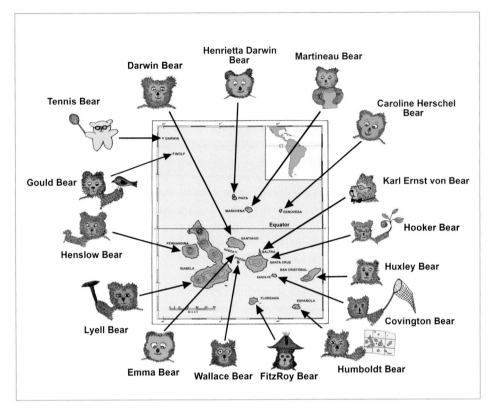

Map of the Galápagos Islands, showing the island on which each bear originally lived.

Harvard University's Edward O. Wilson has recently analyzed certain attributes of the Galápagos bears in conjunction with the specific islands on which each bear originally lived. This discerning scientist has detected two intriguing patterns in these data. The first trend is that the size and elevation of each island appear to be closely related to the height and weight of each bear. Bears from the larger islands, which generally have a mist-drenched highlands, tend to be larger than bears from lower and smaller islands. The smallest bear of all—Tennis Bear—comes from Darwin Island, one of the smallest and most arid islands in the Galápagos group. According to Professor Wilson, the size of a Galápagos bear can be predicted by the equation **S** $= 0.00084 \times$ **H.** (Here **S** represents the size of each bear, and **H** is the height of the island on which the bear once lived.) Hence a 2,975-foot (907 meters) island such as Santiago should have a bear that is about 30 inches (75 cm) tall. By contrast, a low island such as Darwin, which is only 551 feet (168 meters) in elevation, should have a bear only 5.6 inches (14 cm) tall. In actual fact, Darwin Bear (from Santiago) is about 28 inches (70 cm) tall; whereas Tennis Bear (from Darwin Island) is only 3.1 inches (7.8 cm) tall. These facts reinforce the conclusion, first reached by the bears themselves, that each of their different forms evolved over time to adapt them to their homes.

Second, those bears who originally resided on islands closest to Santiago Island (nearly in the center of the archipelago and the home of Darwin Bear), tend to be named after the people to whom Charles Darwin was the most closely attached. For example, Lyell Bear and Hooker Bear, named after two of Darwin's closest friends, both lived on islands that were less than 20 miles (32 km) from Santiago, and Emma Bear (named after Darwin's wife) lived on Rábida Island, right next to Santiago. There is some ongoing debate among Wilson and his colleagues about whether this trend in names is related to a fundamental principle of evolution—namely, the tendency for neighboring forms to be most similar to one another, and, conversely, for the most geographically isolated forms to be the most novel and atypical, as is predicted by Ernst Mayr's 1954 theory about genetic revolutions in small and peripherally isolated populations.

# The Bears' Rules of Tortoise Racing

After leaving the Galápagos Islands with Darwin in 1835, the bears decided to write a 784-page treatise entitled *Rules and Regulations for Tortoise Racing in the Galápagos Islands*. This way, the complete rules of their favorite sport would not be forgotten. A manuscript copy containing these rules exists at Darwin's home in Kent, where it was recently discovered after long having been mistaken for one of Darwin's experiment notebooks. A second copy has been deposited by the bears in the library of the Charles Darwin Research Station.

As the many pages of the *Rules and Regulations* suggest, there are many more rules about tortoise racing than the seven briefly outlined by Darwin Bear in Chapter 3. For further information, one should consult the *Rules and Regulations*, but only on Tuesdays, when there are no tortoise races. As noted earlier, this document has been revised from the original guidelines to reflect the current rules for tourists and visiting scientists. As a result, all tortoise races now involve riderless tortoises selected and trained by the bears.

Besides the rules of tortoise racing, the *Rules and Regulations* contains many other interesting observations about the Galápagos Islands. For this reason, Darwin occasionally cited the *Rules and Regulations* in a work that remained unfinished at the time of his death (see Appendix 3).

# The Untold Story
# of Darwin's Whale-Bear

The passage on page 184 of Darwin's *Origin of Species* (1859)—in which he suggested that a race of aquatic bears might eventually evolve into a creature "as monstrous as a whale"—enjoys particular notoriety among historians of evolutionary biology. As certain aspects of this story are not well known, they deserve to be retold here.

The Galápagos bears were always on the lookout for information about other kinds of bears that might usefully be included in Darwin's *Origin of Species.* One day in 1840, Darwin Bear was reading an old travel book by Samuel Hearne, an explorer of the Canadian wilderness between 1769 and 1772. He found a description of a bear swimming in the water, with his mouth wide open, trying to catch insects. When he mentioned this passage to FitzRoy Bear, FitzRoy got very excited, saying, "Hearne's observation fits in perfectly with my theory about the origin of whales." The two had a lively discussion with Darwin about the possibility that some bearlike animal, living for many generations in an aquatic environment, might have evolved into a marine creature looking like a whale.

FitzRoy Bear had often studied whales during his many sailing trips around the Galápagos Archipelago. He noticed that whales (along with dolphins and porpoises) are fundamentally different from most other sea creatures, because they breathe air and feed their young with mother's milk. Based on this evidence, FitzRoy Bear surmised that whales might actually be very large sea lions, and that

FitzRoy Bear lectures about the relationship between whales and bears.

sea lions might once have been bears. This admittedly speculative theory led the bears to believe that two other Galápagos marine animals (marine iguanas and sea turtles) had evolved from land creatures after they began spending most of their time foraging for food in tidal zones and, later, in the sea.

After much discussion, a brief passage about bears and whales made its way into Darwin's *Origin of Species*. Unfortunately, this particular passage about the whale-bear was seized upon by several of Darwin's critics, some of whom twisted Darwin's words to imply that bears could increase the size of their mouths solely by constant use, rather than by a gradual process of natural selection acting over many generations on small differences in mouth size. Evolution does *not* proceed by changes acquired through use or disuse of a particular organ during a creature's own lifetime, although Darwin himself did endorse a limited role for this putative evolutionary mechanism at a time when the laws of genetics were still poorly understood. Given our much better grasp of the laws of heredity, we now know that there is no way to pass on to one's offspring any of the acquired changes in physical or behavioral traits, in the manner that biologists used to assume.*

Instead, as Darwin realized, the adaptations achieved through evolution are primarily driven by chance variations, such as having a slightly

CHAPTER VI

In North America the black bear was seen by Hearne swimming for hours with widely open mouth, thus catching, like a whale, insects in the water. Even in so extreme a case as this, if the supply of insects we constant, and if better adapted conditions did not already exis the country, I can see no diff in a race of bears being ren by natural selection, more a

184

Following FitzRoy's lecture about bears and whales, Darwin Bear drafts a key passage for Darwin's possible inclusion in the *Origin of Species*.

larger or smaller nose, mouth, tooth, arm, or leg. If these chance variations are adaptive and heritable, they tend to be naturally selected and preserved over subsequent generations. For example, giraffes evolved their long necks not by stretching them but because giraffes that happened to be born with longer necks were able to reach higher into treetops for food and thus were better able to survive when vegetation was scarce. Their offspring, who tended to inherit their parents' trait of having a long neck, also prospered.

---

* Beliefs about the inheritance of acquired characteristics were previously essential to the now rejected theory of evolution associated with the French biologist Jean-Baptiste Lamarck, and for this reason they are sometimes called "Lamarckism." On a much more limited scale, modern science has shown that certain developmental experiences (including those that occur before birth) sometimes cause a rewriting of DNA, which can then be inherited by the next generation. See further Kevin V. Morris, "Lamarck and the Missing Lnc," *The Scientist* (October 1, 2012).

In Darwin's day, people sometimes confused his theory of evolution with the theory of change through "use and disuse," and Darwin's own limited endorsement of the latter theory inevitably added to this confusion. It was therefore unfortunate that Darwin's whale-bear story, contrary to his intention, ended up undermining the core message of his book: namely, that natural selection is the primary driver of adaptive evolutionary change. "It is laughable," Darwin wrote to a correspondent about this misunderstanding in 1861, "how often I have been attacked and misrepresented about this bear."

To understand the subsequent fate of Darwin's discussion about the whale-bear, it is worth reconsidering the original passage from the *Origin of Species*, which reads as follows:

> In North America the black bear was seen by Hearne swimming for hours with widely open mouth, thus catching, like a whale, insects in the water. Even in so extreme a case as this, . . . I can see no difficulty in a race of bears being rendered, by natural selection, more and more aquatic in their structure and habits, with larger and larger mouths, till a creature was produced as monstrous as a whale. (*Origin of Species*, p. 184)

In the second edition of the *Origin of Species*, published in 1860, Darwin tried to deflect some of the criticisms associated with this controversial passage by inserting the word "almost" before the phrase "like a whale," hoping that his comparison between bears and whales would not be taken so literally. On the advice of the comparative anatomist and paleontologist Richard Owen, Darwin also decided to remove the entire second sentence from the book, although he always regretted having done so, as he told R. G. Whiteman two decades later.

Even Richard Owen, who was one of Darwin's harshest critics, seems to have agreed privately with Darwin about the close relationship between whales and bears. In a conversation with Darwin in

1859, Owen confessed: "I was more struck with this than any other passage [in the *Origin of Species*]; you little know of the remarkable & essential relationship between bears & whales." Yet in his subsequent review of the *Origin of Species,* Owen criticized this very passage for what he considered its wildly speculative nature!

The fact that Owen could bring himself to ridicule Darwin in public for an idea he apparently believed to be true in private is one of the reasons why Darwin lost virtually all respect for Owen after 1859. It is also one of the reasons why no Galápagos bear ever wanted to be named after Owen, who was arguably the most famous British biologist at the time Darwin's *Origin of Species* was published, and whose fame faded owing to his non-Darwinian way of thinking. However, the bears did sometimes use the phrase "Owen Bear"—as in "Don't be such an Owen Bear"—whenever one of them would do something annoying.

We owe it to the Galápagos bears that the story about the whale-bear did not disappear from *all* later editions of the *Origin of Species.* Darwin had already deleted this passage in December 1859 from the soon-to-be published second British edition. But Darwin Bear prevailed upon him to allow it to remain in the "New Edition Revised and Augmented by the Author" that was published by Appleton in America in July 1860. "Let's try an experiment," Darwin Bear suggested. "There are many more bears in North America than in Britain. For this reason, people living on this continent may be much friendlier to bears and perhaps more accepting, as well, of the idea of bears turning into whales. So let's keep the whale-bear passage in the new American edition of the *Origin of Species.*" Because Appleton kept reprinting this American edition until 1870, when the publisher sought Darwin's permission to reprint the text of the fifth British edition, American readers experienced a full decade of exposure to Darwin's whale-bear theory.

Today, of course, almost no one doubts that Darwin was right, in the broadest sense, about the ability of natural selection to create a

A 47-million-year-old ancestor of whales (*Rodhocetus*), recently discovered in fossilized form in Pakistan. The hands and feet appear to have been webbed. This creature combined the ankle bones of sheep, deer, and hippopotami with the skull bones of an archaic whale. (After a painting by John Klausmeyer in *Science* [2001])

whale-like creature from a land mammal resembling a bear. Whales, dolphins, and porpoises are all now known to have evolved from a four-legged land mammal that first took to the sea around 60 million years ago, some 5 million years after the last dinosaurs roamed the earth. However, it is also now known from fossil evidence that bears are not *directly* related to whales, because at least two different groups of land mammals began to pursue life in the ocean around this time. One group, most closely allied to cows and hippos, evolved into cetaceans (including whales) and another group, much more closely related to bears, evolved into the pinnipeds (sea lions, seals, and walruses). Of course, going back even further in time, bears and whales do share a common ancestor, as do all mammals. It is rumored that one of Darwin's manuscripts, unfinished at his death in 1882 and later lost, described these relationships in greater detail. This unfinished book, to which all of the Galápagos bears contributed ideas and suggestions, was entitled *The Descent of Bears: A Whale of a Tale.*

Darwin was right that a race of land mammals, given enough time for adaptation to life in an aquatic environment, might evolve into something "as monstrous as a whale." What Darwin's controversial whale-bear story tells us, however, is that this idea was far too revolutionary in 1859 even for one of the most revolutionary of all scientific books. And this is why later editions of the *Origin of Species*—a book that started out as being all about bears—say remarkably little about them.

# The Bears' Recipe for
# Darwinberry Jam

The bears have adapted their recipe from a gooseberry jam recipe given to them by Darwin's wife, Emma:

> In a medium-sized, stainless-steel saucepan, mix 4 cups of blackberries, ¼ cup of blueberries, ¼ cup of lime juice and 2 dashes of Angostura bitters, 2 or 3 pieces of finely chopped lemon rind, a pinch of cinnamon, 2 ground coffee beans, ¼ cup of lemon juice, and ¼ cup of water. Stir over a medium flame. Once this mixture comes to a boil, add approximately 4 cups of sugar (or ¾ cup for each cup of fruit and liquid mixture). Stir frequently with a wooden spoon to keep the mixture from sticking to the pan. Boil until the mixture has the consistency of jam (about 30 minutes). Pack the jam into sterilized jars, seal, and store in a cool place.

For Darwinberry Jam Deluxe, substitute an equal quantity of blueberries for blackberries.

By cooking a portion of this Darwinberry mixture until it reaches the consistency of hard candy (at about 300°F [150°C]), one can make what the bears call "Conservation Pellets." While still hot to very warm to the touch, roll this candy mixture into round spheres, each about 1 inch (2.5 cm) in diameter. Then allow them to cool. When walking around in the Galápagos Islands, carry these hard candy balls with you and throw them at any feral goats or pigs that are eating the native vegetation. During a tortoise race, scoring a direct hit to a feral animal with a Darwinberry pellet earns a contestant a 20-second reduction in racing time (see Rule 245 in the *Rules and Regulations for Tortoise Racing in the Galápagos Islands*). After the recent revision of the *Rules and Regulations* to accord with Galápagos National Park guidelines, Conservation Pellets are now thrown by each bear as they walk alongside a champion tortoise along the racecourse. Tennis Bear uses specially made tiny pellets, which, by swatting them with his racket, he is able to direct toward his targets with remarkable speed and accuracy.

# Are There Really Any Bears in the Galápagos Islands?

Prior to publication of the Darwin Bear story, which has revealed to the world a new and uniquely Galápagos genus of bears, scientists had not encountered evidence of any bears in this archipelago. Of course, the absence of historical evidence that bears were once present in these islands can be explained by the unusual circumstances, recounted here, of their imminent extinction in the 1830s, and the fact that the sixteen surviving bears were all taken back to England by Darwin aboard H.M.S. *Beagle*. As we have learned in *Darwin and His Bears*, it is because evidence of bears in the Galápagos could not be corroborated that Darwin himself decided not to mention the bears in the *Origin of Species*. Still, a scientific skeptic might be justified in concluding that there never were any bears in the Galápagos and that *Darwin and His Bears* is merely a work of fiction. So how can we know the truth?

Fortunately, we can apply the scientific method in our aim to reach a definitive answer to this question. The essence of the scientific method lies in the systematic effort to test factual claims and hypotheses by gathering relevant evidence, and this approach sometimes requires conducting carefully designed experiments, just as Darwin did in order to further his *Origin of Species* argument. Darwin often tested ideas and theories that most of his scientific peers disagreed with, only to discover that these colleagues (and conventional wisdom) were wrong. "I love fools' experiments," he once said, adding, "I am always making them."

Darwin also understood that scientists need to be especially careful not to dismiss evidence that contradicts their current beliefs. Facing up to contrary evidence is often difficult, because we tend to disregard information that conflicts with our expectations. This was why Darwin observed what he described in his *Autobiography* as a "golden rule"—namely, to record immediately any facts that seem contrary to one's theories. For example, a scientist who is already convinced that bears do not exist in the Galápagos, but who happens to catch a fleeting glimpse of one at dusk near the Charles Darwin Research Station library (where they are purported to sleep at night), might well decide that he or she has seen a feral goat rather than a bear and therefore forget the whole matter. By contrast, after failing to see any bears for months, or even years, someone who believes that bears *do* inhabit the Galápagos might conclude that they were simply elsewhere during the period of observation. The key to good science is to devise formal tests that overcome the human bias for preferring evidence that confirms one's expectations. Unfortunately, devising such definitive tests is not always easy.

The following suggestions represent some of the ways in which one might go about testing the claim that bears now live, or have previously lived, in the Galápagos Islands.

1.  Tourists visiting the Galápagos might want to be on the lookout for the bears, who may be doing fieldwork on one or more of the thirteen islands where visitors are permitted to land. Any compelling photographic evidence of bears encountered during such visits can be posted on this book's website, darwinbear.com. Photographs should be identified by island as well as by specific visitor site, and, if possible, should include geographically recognizable features in the background. Photos should also be dated to facilitate tracking the bears' movements in the archipelago.

2. After obtaining permission from the Galápagos National Park Directorate, one could place large bowls of blueberries in selected places to see if any bears show up. These sites would ideally be recorded photographically, day and night, with camera traps.

3. With permission of the Darwin Station librarian, one might enter the library late at night to see if bears really do sleep there in the rafters, as it is claimed they do. An even more thorough monitoring of the library might be achieved with the use of camera traps set to detect nighttime entry into the building.

4. A relatively new technique called environmental DNA (or eDNA) offers considerable promise in settling the question of whether there are any bears in the Galápagos. Scientists who study particularly elusive species often test for their presence by collecting DNA samples from the environment. If such DNA samples from the Galápagos were to turn up genetic material assignable to the Ursidae (bear family), this would be compelling evidence.

5. One could advertise a tennis tournament in the monthly Galápagos publication *El Colono* and see if a small bear with a tennis racket shows up.

Interested readers of this book are encouraged to think of other ways that might prove, one way or the other, whether any bears live in the Galápagos Islands. People who believe they have particularly good suggestions about how to detect the bears are invited to post their ideas on this book's website, darwinbear.com.

Although the following evidence does not provide a formal scientific test of the hypothesis that bears have lived in the Galápagos, it is nevertheless indirectly relevant. First we must ask a question: How could bears possibly have reached the Galápagos from their original home—presumably the South American mainland? In the

*Origin of Species*, Darwin himself considered this question of how distant oceanic islands had been colonized by animals and plants; he felt it to be one of the most serious potential objections to his evolutionary theories. For example, land snails are commonly found on oceanic islands, but they die quickly when submerged in salt water, which seems to refute the explanation that they arrived on islands by rafting on floating vegetation. After months of experimentation, Darwin discovered that land snails sometimes estivate to avoid desiccation. When in a state of estivation, snails seal off the opening of their shells with a thin layer of mucus that dries in the air. In such a state, as Darwin found to his considerable relief, land snails can survive in salt water for more than enough time needed for ocean currents to carry them for thousands of miles.

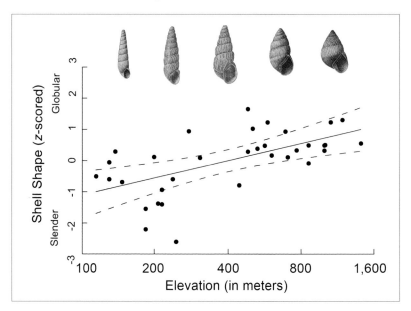

Land snails living in the elevated part of islands, including *Bulimulus darwini*, have relatively large apertures relative to their rounded shells. By contrast, land snails residing in the lowlands have much smaller apertures and elongated shells, features that help these land snails to retain moisture in an arid environment. Figure adapted from data in Robert P. Smith (1971) and figures in William Healey Dall (1896).

A Galápagos land snail (*Bulimulus darwini*), depicted on a native orchid (*Ionopsis utricularioides*). This endemic species of land snail was first collected by Darwin in the highlands on Santiago in October 1835. Having survived transport to the Galápagos Islands several million years ago—most likely by rafting on floating vegetation—the ancestor of all present-day Galápagos land snails subsequently evolved into more than eighty recognized species and subspecies, making this group the champion "speciator" and unsung evolutionary hero in these islands. More than half of these species are now either extinct or critically endangered owing to habitat loss and the effects of introduced species.

A rice rat (*Aegialomys galapagoensis*) on Santa Fé Island. These nocturnal mammals are remarkably tame and acrobatic. Once they have discovered a researcher's campsite, half a dozen or more will visit each evening as they forage for food. They have no difficulty jumping nearly a meter off the ground onto any object that might contain something edible, and they are almost impossible to deter from their explorations. At night it is necessary to leave one's tent open, as otherwise rice rats will chew through the tent cloth to get in. I have been awakened many times on Santa Fé to find one or more rice rats exploring my tent, sometimes brushing their whiskers against my face as they look for food. Tameness is a characteristic of almost all native Galápagos species and has evolved owing to the isolation of the islands from various predatory species, including humans. When fear becomes less adaptive, boldness and curiosity are generally favored by natural selection.

There is another indirect source of support for the existence of bears in the Galápagos—one closely related to Darwin's findings about long-distance rafting. The Galápagos Islands are currently home to an animal that holds the world record for the largest distance ever crossed over open ocean by a terrestrial mammal. This animal is the endemic rice rat, which is represented by several surviving species and subspecies (unfortunately, some local populations of this taxon are now extinct owing to the introduction of the Black and Norwegian rats in historical times). If rice rats could make the perilous journey to the Galápagos—by rafting across 600 miles (1,000 km) of ocean—then small bears might have been able to do so too. In addition, a paleontologist has found that these islands were once home to a giant rat (*Megaoryzomys curioi*), which was as big as a cat (and hence even larger than some of the Galápagos bears). It is believed that this giant rat also arrived in the islands by rafting. This particular rat species is currently known only from fossil evidence (Steadman and Ray, 1982), having become extinct sometime after the arrival of humans in the Galápagos.

Yet another source of evidence favoring the existence of bears in the Galápagos is the fact that Darwin Bear and his assistant Covington Bear are believed to have attended a recent workshop for the Galápagos Verde 2050 Project. This ambitious project is trying to restore the islands to their conditions before the introduction of feral animals and plants. Following the workshop, Darwin Bear and Covington Bear were both given certificates of attendance (reproduced on p. 154), signed by the director of the Charles Darwin Research Station and by the head of the Galápagos Verde 2050 Project.

By considering suggestions for how to test for the existence of Galápagos bears, readers should bear in mind—as the well-known aphorism goes—that absence of evidence is not necessarily evidence of absence. This is to say, new species of Galápagos plants, insects, reptiles, and birds have been discovered in the Galápagos even after

Certificates of attendance by Darwin Bear ("Darwin Osito," in Spanish) and Covington Bear at a Galápagos Verde 2050 Workshop held at the Charles Darwin Research Station on Santa Cruz Island on October 23–27, 2017. The certificate is signed by Arturo Izurieta, director of the Charles Darwin Research Station, and Patricia Jaramillo, head of the Galápagos Verde 2050 Project for ecological restoration of the islands.

two centuries of systematic collecting and research. As a good example, until just recently it was believed that only two species of land iguanas live in the Galápagos. But in 1986, more than a hundred and fifty years after Darwin's visit, a rare pink-and-black striped iguana was found living on Volcán Wolf, the northernmost volcano of Isabela. We must accept the fact that scientific evidence is always provisional, and hence it may change based on new information. Moreover, if the Galápagos bears are secretive, as they are purported to be, they may not *want* to be found by scientists, making proof of their existence even more difficult.

———— **APPENDIX 6** ————

# Why Does the Darwin Bear Story Have Six Appendices?

If humans have only one appendix, it may be asked, "Why does *Darwin and His Bears* have six of them?" The answer to this question is simple. The human appendix was long, but incorrectly, thought to be a "rudimentary" or "vestigial" organ—one that was once supposedly much larger than it now is and that formerly played a greater role in digestion at a time when our hominid ancestors' diet contained a larger portion of vegetable matter.

Today, scientific evidence has shown that the human appendix is not entirely vestigial, contrary to what Darwin argued in the *Descent of Man* (1871), but rather is a fully functional organ that plays an important role in aiding the body's defenses against bacteria and other sources of disease. The appendix, it turns out, helps our bodies by trapping these foreign materials and producing antibodies against them. (The tonsils at the back of our throat serve a similar germ-collecting function.)

This recent scientific reinterpretation of the role of the human appendix accords nicely with the fact that *Darwin and His Bears* has six appendices. Like the contents of every digestive system, some books have extensive material in them that really should not be there. The six appendices of *Darwin and His Bears* serve the valuable function of keeping this material out of the main body of the book, thus protecting readers from the alluring distraction of tangential material.

# Further Reading

**Books about Darwin's life and scientific work in general**

For a superb biography of Darwin, see Janet Browne, *Charles Darwin: Voyaging. Volume I of a Biography* (New York: Alfred A. Knopf, 1995); and *Charles Darwin: The Power of Place. Volume II of a Biography* (New York: Alfred A. Knopf, 2002). Also recommended is Sir Gavin de Beer's *Charles Darwin: A Scientific Biography* (Garden City, NY: Doubleday, 1965). For a useful encyclopedic approach to Darwin, see Paul van Helvert and John van Wyhe's *Darwin: A Companion* (Singapore: World Scientific, 2020).

*The Autobiography of Charles Darwin*, written in 1876 and published as an expurgated text after his death in 1882, also supplies an excellent introduction to Darwin's life and work (*Life and Letters of Charles Darwin, including an Autobiographical Chapter*, 3 vols., Francis Darwin, ed. [London: John Murray, 1887]). Nora Barlow's edition of the *Autobiography* (New York: W. W. Norton, 1958) has restored the omissions that were considered too sensitive for publication in the 1887 version, including some of Darwin's discussions about religion. At the bears' request, Darwin's *Autobiography* omits the influential part the sixteen Galápagos bears played in his life. Darwin also omitted any reference to the bears' blueberries in his *Autobiography*, fearing that revelations about these berries might endanger the bears' and his own family's desire for privacy, especially given how delicious the berries were and the expectation that everyone would want to try them. Blueberries were finally introduced for commercial sale in England in the 1930s.

Darwin's own account of his five-year circumnavigation of the world was published as *Journal of Researches into the Geology and Natural*

*History of the Various Countries Visited by H.M.S. "Beagle," under the Command of Captain FitzRoy, R.N., from 1832 to 1836* (London: Henry Colburn, 1839). This work provides a fascinating and eminently readable account of the crucial years that Darwin spent circumnavigating the globe and has never been out of print. A second edition, published in 1845 by John Murray, contains considerable additions to the chapter about Darwin's Galápagos visit. The book is currently available in many modern editions, including in Edward O. Wilson, ed., *From So Simple a Beginning: Darwin's Four Great Books,* 4 vols. (New York: W. W. Norton, 2010), which also includes Darwin's *Origin of Species* (1859), *Descent of Man* (1871), and *Expression of Emotions in Man and Animals* (1872).

*The Annotated Origin: A Facsimile of the First Edition*, edited by James T. Costa, supplies an invaluable introduction to Darwin's *Origin of Species* (Cambridge, MA: Harvard University Press, 2009). Costa has also published a fascinatingly detailed account of Darwin's experimental work and its contribution to the *Origin of Species* and other publications (*Darwin's Backyard: How Small Experiments Led to a Big Theory* (New York: W. W. Norton, 2017).

Two additional sources on Darwin's scientific thought are particularly useful. Michael T. Ghiselin's *The Triumph of the Darwinian Method* (Berkeley, CA: University of California Press, 1969) contains a compelling analysis of how all of Darwin's works, even his post-*Origin* works in botany, were part of a unified effort to explain evolutionary processes. Ghiselin also shows that Darwin, in spite of his claims to have worked by true Baconian induction, was in fact guided by the much more sophisticated hypothetico-deductive method and, in this respect, was considerably ahead of his time. An equally cogent argument for the unified nature of Darwin's scientific enterprise may be found in Duncan M. Porter and Peter W. Graham, *Darwin's Sciences: How Charles Darwin Voyaged from Rocks to Worms in His Search for Facts to Explain How the Earth, Its Geological Features, and Its Inhabitants Evolved* (Oxford: Wiley Blackwell, 2016).

## Chapter 1 and Appendix 3

On the origins of whales (and their relationship to bears), see Stephen Jay Gould, "Hooking Leviathan by Its Past," in *Dinosaur in a Haystack* (New York: Harmony Books, 1995), pp. 359–76; and, for a more technical treatment, P. D. Gingerich, M. Haq, I. S. Zalmout, I. H. Khan, and M. S. Malkani, "Origin of Whales from Early Artiodactyls: Hands and Feet of Eocene Protocetidae from Pakistan," *Science* 293 (2001), 2239–42.

## Chapters 2–4

For information on the role that the Galápagos Islands played in the development of Darwin's evolutionary thinking, see Frank J. Sulloway, "Darwin and His Finches: The Evolution of a Legend," *Journal of the History of Biology* 15 (1982), 1–53; "Darwin's Conversion: The *Beagle* Voyage and Its Aftermath," *Journal of the History of Biology* 15 (1982), 325–96; "The *Beagle* Collections of Darwin's Finches (*Geospizinae*), *Bulletin of the British Museum (Natural History) Zoology Series* 43, no. 2 (1982); "Darwin and the Galapagos," *Biological Journal of the Linnean Society* 21 (1984), 29–59; and "Tantalizing Tortoises and the Darwin-Galápagos Legend," *Journal of the History of Biology* 42 (2009), 3–31. Based on previously unknown archival sources as well as examination of Darwin's type specimens at the British Museum, these publications show that Darwin did not label his Galápagos birds by island and that he later sought to remedy his collecting oversights by obtaining information about the island localities of the Galápagos birds shot by three other *Beagle* collectors. These revisionist historical accounts also debunk the myth that Darwin's famous Galápagos finches inspired his theory of evolution by natural selection, showing instead how ornithologist John Gould corrected Darwin's initially mistaken belief that these birds were members of four different avian families—a key rectification that paved the way for Darwin's belated evolutionary understanding of this evidence. Unfortunately, these various publications—written before the author met Darwin

Bear—omit the important part played by the Galápagos bears in Darwin's intellectual development.

Evidence for a split between the land and marine iguana lineages 10 to 20 million years ago is presented in Kornelia Rassmann, "Evolutionary Age of the Galápagos Iguanas Predates the Age of the Present Galápagos Islands," *Molecular Phylogenetics and Evolution* 7 (1997), 158–72. Because the oldest present island in the Galápagos group has been dated to about 4.2 million years, these data support the existence of sunken islands on which the ancestral iguana lineage formerly lived before it split into the land and marine forms. See also K. Rassman, D. Tautz, F. Trillmich, and C. Gliddon, "The Microevolution of the Galápagos Marine Iguana *Amblyrhynchus cristatus* Assessed by Nuclear and Mitochondrial Genetic Analysis," *Molecular Ecology* 6 (1997), 437–52.

## Chapter 5

Since the initial publications on Darwin's finches by John Gould and Darwin in 1837, these birds have been the subject of considerable additional research. David Lack's classic book *Darwin's Finches* (Cambridge: Cambridge University Press, 1947) is still very much worth reading despite many new and important discoveries about those birds since its publication. For an account of more than four decades of painstaking research documenting the year-by-year ebb and flow of natural selection in Darwin's finches, see Peter R. Grant and B. Rosemary Grant's *How and Why Species Multiply: The Radiation of Darwin's Finches* (Princeton, NJ: Princeton University Press, 2008). For molecular genetic analyses of Darwin's finches, which reveal repeated episodes of hybridization and hence the difficulty of identifying individual species lineages among these birds, see Heather L. Farrington, Lucinda P. Lawson, Courtney M. Clark, and Kenneth Petren, "The Evolutionary History of Darwin's Finches: Speciation, Gene Flow, and Introgression in a Fragmented Landscape," *Evolution* 68 (2014), 2932–44; and Sangeet Lamichhaney, Jonas Berglund,

Markus Sällman Almén, Khurram Maqbool, Manfred Grabherr, Alvaro Martinez-Barrio, Marta Promerova, Carl-Johan Rubin, Chao Wang, Neda Zamani, B. Rosemary Grant, Peter R. Grant, Matthew T. Webster, and Leif Andersson, "Evolution of Darwin's Finches and Their Beaks Revealed by Genome Sequencing," *Nature* 518 (2015), 371–75. An engrossing overview of the four decades of pathbreaking research by the Grants is provided in Jonathan Weiner's Pulitzer Prize–winning *The Beak of the Finch: A Story of Evolution in Our Time* (New York: Alfred A. Knopf, 1994).

For research showing the rapidity with which Darwin's finches can evolve in response to environmental changes and chance events, see Sangeet Lamichhaney, Fan Han, Matthew T. Webster, Leif Andersson, B. Rosemary Grant, and Peter R. Grant, "Rapid Hybrid Speciation in Darwin's Finches," *Science* 359 (2018), 224–28; and Frank J. Sulloway and Sonia Kleindorfer, "Adaptive Divergence in Darwin's Small Ground Finch (*Geospiza fuliginosa*): Divergent Selection along a Cline," *Biological Journal of the Linnean Society* 110 (2013), 45–59. These iconic birds are now seriously threatened by an introduced ectoparasite (*Philornis downsi*), whose larvae feed on nestlings and cause upward of 90 percent mortality in some years; see Jody A. O'Connor, Frank J. Sulloway, Jeremy Robertson, and Sonia Kleindorfer, "*Philornis downsi* Parasitism Is the Primary Cause of Nestling Mortality in the Critically Endangered Medium Tree Finch (*Camarhynchus pauper*)," *Biodiversity and Conservation* 19 (2010), 853–66; Sonia Kleindorfer, Jody A. O'Connor, Rachel Y. Dudaniec, Steven A. Myers, Jeremy Robertson, and Frank J. Sulloway, "Species Collapse via Hybridization in Darwin's Tree Finches," *American Naturalist* 183 (2014), 325–41; and Sonia Kleindorfer and Frank J. Sulloway, "Naris Deformation in Darwin's Finches: Experimental and Historical Evidence for a Post-1960s Arrival of the Parasite *Philornis downsi*," *Global Ecology and Conservation* 7 (2016), 122–31.

A useful reference for the Galápagos fauna as a whole is Andy Swash and Rob Still, *Birds, Mammals, and Reptiles of the Galápagos*

*Islands: An Identification Guide* (New Haven and London: Yale University Press, 2005). For a guide to the flora of the Galápagos, see Conley K. McMullen, *Flowering Plants of the Galápagos* (Ithaca, NY, and London: Cornell University Press, 1999); and Ira L. Wiggins and Duncan M. Porter, *Flora of the Galápagos Islands* (Stanford, CA: Stanford University Press, 1971).

On science as a social process, see Philip Kitcher, *The Advancement of Science: Science Without Legend, Objectivity Without Illusions* (Oxford: Oxford University Press, 1993); David Hull, *Science as a Process: An Evolutionary Account of the Development of Science* (Chicago: Chicago University Press, 1988); Bruno Latour and Steven Woolgar, *Laboratory Life: The Construction of Scientific Facts*, 2nd ed. (Princeton, NJ: Princeton University Press, 1986) and Steven Shapin and Simon Schaffer, *Leviathan and the Air Pump* (Princeton, NJ: Princeton University Press, 1985).

## Chapter 6

The famous debate between Thomas Henry Huxley and Bishop Samuel Wilberforce took on legendary proportions as the years went by, especially in the recollections of the Darwinians who were there that day. Darwin Bear's own account of this episode reflects some of the triumphant sentiments that have colored other retrospective accounts. For further information on this debate, and how historical interpretations of it have changed over time, see Edward Caudill, *Darwinian Myths: The Legends and Misuses of a Theory* (Knoxville: University of Tennessee Press, 1997); and Ian Hesketh, *Of Apes and Ancestors: Evolution, Christianity, and the Oxford Debate* (Toronto: University of Toronto Press, 2012).

## Chapter 7

The question of why Darwin did not publish the *Origin of Species* until 1859, twenty-one years after his discovery of natural selection, is ably examined by John van Wyhe in "Mind the Gap: Did Darwin Avoid

Publishing His Theory for Many Years?," *Notes and Records* 61 (2007), 177–208. In particular, van Wyhe refutes the myth that Darwin delayed publication of his controversial book because he feared the intense criticism it would unleash. Rather, Darwin believed it was incumbent upon him to publish first all of the research growing out of the *Beagle* voyage. These scientific findings, which were set forth in five books (issued in twelve volumes) as well as in sixteen shorter publications, appeared between 1837 and 1854. Only in late 1854 did Darwin finally consider it appropriate to begin drafting what he called his "Big Book," to which he gave the working title *Natural Selection*. Unpublished in Darwin's lifetime, this work finally appeared in 1975 as *Charles Darwin's Natural Selection: Being the Second Part of His Big Species Book Written from 1856 to 1858*, R. C. Stauffer, ed. (Cambridge: Cambridge University Press). This big book was then reduced to a much shorter work, the *Origin of Species,* after Alfred Russel Wallace anticipated the theory of natural selection in 1858 and Darwin felt compelled to get something quickly into print. It should be noted that this controversial work benefited greatly from the additional research, experiments, and thinking that Darwin did as he was preparing his various *Beagle* researches for publication.

Some of the most important misconceptions of "psicko-analytic" theory are detailed in Frank J. Sulloway, *Freud, Biologist of the Mind: Beyond the Psychoanalytic Legend* (New York: Basic Books, 1979; Cambridge, MA: Harvard University Press, 1992). This book also documents the extensive influence exerted on Freud's thinking by Ernst Haeckel's biogenetic law (the erroneous but widely influential nineteenth-century theory that ontogeny recapitulates phylogeny). See also "Psychoanalysis and Pseudoscience: Frank J. Sulloway Revisits Freud and His Legacy," in Todd Dufresne, ed., *Against Freud: Critics Talk Back* (Stanford, CA: Stanford University Press, 2007), pp. 48–69. Another useful source of the failings of psychoanalysis is Malcolm Macmillan's *Freud Evaluated: The Completed Arc* (Cambridge, MA: MIT Press, 1997). For Freud's failings as a clinician, see Frank J. Sulloway,

"Reassessing Freud's Case Histories: The Social Construction of Psychoanalysis," *Isis* 82 (1991), 245–75; and especially Frederick Crews, *Freud: The Making of an Illusion* (New York: Metropolitan Books, 2017).

The extensive influence that Ernst Haeckel's biogenetic law had on late-nineteenth-century scientific thought, and outside science as well, is ably documented in Stephen Jay Gould, *Ontogeny and Phylogeny* (Cambridge, MA,: Belknap Press of Harvard University Press, 1977).

## Chapter 9

Research showing that fruit color is an honest signal of anthocyanins (antioxidant pigments that promote health) is described in H. M. Schaefer, K. McGraw, and C. Catoni, "Birds Use Fruit Colour as Honest Signal of Dietary Antioxidant Rewards," *Functional Ecology* 22 (2008), 303–10. Ernst Mayr's pioneering views on genetic revolutions in isolated populations are set forth in "Change of Genetic Environment and Evolution," in Julian Huxley, A. C. Hardy, and E. B. Ford, eds., *Evolution as a Process* (London: Allen and Unwin, 1954), pp. 157–80. The theory of punctuated equilibrium is presented by Niles Eldredge and Stephen Jay Gould in "Punctuated Equilibria: An Alternative to Phylogenetic Gradualism," in Thomas J. M. Schopf, ed., *Models in Paleobiology* (San Francisco: Freeman Cooper, 1972), pp. 82–115.

Edward O. Wilson's influential theories about island biogeography are set forth in Robert H. MacArthur and Edward O. Wilson, *The Theory of Island Biogeography* (Princeton, NJ: Princeton University Press, 1967). MacArthur, who died of renal cancer at the age of forty-two, also played an important role in the development of niche partitioning theory and theoretical ecology. See Robert H. MacArthur, *Geographical Ecology: Patterns in the Distribution of Species* (New York: Harper & Row, 1972); and Stephen D. Fretwell, "The Impact of Robert MacArthur on Ecology," *Annual Review of Ecology and Systematics* 6 (1975), 1–13. Wilson's sweeping overview of human knowledge and how to improve it is presented in *Consilience: The Unity of Knowledge* (New York: Alfred A. Knopf, 1998). The term "consilience" was coined by William Whewell in *The Philosophy of the Inductive Sciences, Founded on Their History* (London: John W. Parker, 1840). Whewell derived this term from the Latin *con* (together) and *siliens* (jumping) to denote how multiple sources of independent evidence can lead to stronger conclusions.

## Chapters 10–11

The problem of introduced species in the Galápagos and other sources of ecological damage caused by human habitation is treated in Michael D'Orso's *Plundering Paradise: The Hand of Man on the Galápagos Islands* (New York: HarperCollins, 2002). A good overview of the invasive-species problem in the Galápagos is also provided by M. Verónica Toral-Granda, Charlotte E. Causton, Heinke Jäger, Mandy Trueman, Juan Carlos Izurieta, Eddy Araujo, Marilyn Cruz, Kerstin K. Zander, Arturo Izurieta, and Stephen T. Garnett, "Alien Species Pathways to the Galapagos Islands, Ecuador," *PLOS ONE* 12 (2017): e0184379. For an example of an ecological cascade caused by the extermination of the Galápagos hawk on the island of Santa Cruz and its effect on the giant tree *Opuntia* population on Plaza Sur, see Frank J. Sulloway, "The Mystery of the Disappearing *Opuntia*," *Galápagos Matters*, Autumn/Winter (2015), 8–9.

Two excellent reviews of the history of scientific research in the Galápagos are Edward J. Larson, *Evolution's Workshop: God and Science on the Galapagos Islands* (New York: Basic Books, 2001); and Matthew J. James, *Collecting Evolution: The Galapagos Expedition That Vindicated Darwin* (New York: Oxford University Press, 2017). For an engrossing history of the Galápagos that recounts some of the human tragedies that have been associated with it, see Octavio Latorre, *The Curse of the Giant Tortoise: Tragedies, Crimes and Mysteries in the Galápagos Islands*, 3rd ed. (Quito: Artes Gráficas Señal, 1999).

## Appendix 5

The tendency of the human mind to seek out confirming evidence and to reject conflicting evidence is ably treated in Michael Shermer,

*The Believing Brain: From Ghosts and God to Politics and Conspiracies—
How We Construct Beliefs and Reinforce Them as Truths* (New York:
Henry Holt, 2011).

The ability to extract and analyze DNA from environmental samples constitutes a major scientific breakthrough for the monitoring
of species, populations, and ecological communities and has proved
particularly useful in conservation work (Philip Francis Thomsen
and Eske Willerslev, "Environmental DNA—An Emerging Tool in
Conservation for Monitoring Past and Present Biodiversity," *Biological Conservation* 183 [2015], 4–18). Such eDNA studies include the
analysis of ancient DNA found in permafrost, tundra, glaciers, and
soil sediments, as well as modern DNA extracted from oceans, lakes,
and terrestrial environments. Use of eDNA has helped to overcome
some of the problems associated with other monitoring techniques.
These problems include incorrect identification of cryptic species,
insufficient taxonomic expertise among observers, inconsistencies
between different sampling methods, and the fact that traditional
survey methods can be disruptive to the populations and environments under study.

Evidence of vertebrate fossil taxa from the Galápagos, found in
lava tubes and caves, is provided by David W. Steadman, *Holocene Vertebrate Fossils from Isla Floreana, Galápagos* (*Smithsonian Contributions
to Zoology* 413 [1986]); David W. Steadman and Clayton E. Ray, *The
Relationships of* Megaoryzomys curioi, *an Extinct Cricetine Rodent (Mur-
oidea, Muridae) from the Galápagos Islands, Ecuador* (*Smithsonian Contributions to Paleobiology* 51 [1982]); and David W. Steadman, Thomas
W. Stafford, Jr., Douglas J. Donahue, and A. J. T Jull, "Chronology
of Holocene Vertebrate Extinction in the Galápagos Islands," *Qua-
ternary Research* 36 (1991), 126–33. Information on the discovery of
living descendants of Floreana tortoises, long thought to be extinct,
may be found in Nikos Poulakakis, Scott Glaberman, Michael Rus-
sello, Luciano B. Beheregaray, Claudio Ciofi, Jeffrey R. Powell, and
Adalgisa Caccone, "Historical DNA Analysis Reveals Living Descen-

dants of an Extinct Species of Galápagos Tortoise," *Proceedings of the National Academy of Sciences* 105 (2008), 15464–69. On the 1986 discovery of a new species of land iguana, see Gabriele Gentile, Anna Fabiani, Marquez Cruz, Howard Snell, Heidi M. Snell, Washington Tapia, and Valerio Sbordoni, "An Overlooked Pink Species of Land Iguana in the Galápagos," *Proceedings of the National Academy of Sciences* 106 (2009), 507–11.

For the impressive evolutionary story about Galápagos land snails, see Christine E. Parent and Bernard J. Crespi, "Sequential Colonization and Diversification of Galápagos Endemic Land Snail Genus *Bulimulus* (Gastropoda, Stylommatophora)," *Evolution* 60 (2006), 2311–28. The relationship between land-snail shell shape and habitat elevation is analyzed by Robert P. Smith in "Factors Governing the Dispersal, Distribution, and Microevolution of Land Snails of *Naesiotus* in the Galapagos Archipelago," B.A. honors thesis, Harvard University, Cambridge, MA (1971). Illustrations of the five *Bulimulus* species in Appendix 5, which show the relationship between shell shape and habitat elevation, are reproduced from William Healy Dall, "Insular Landshell Faunas, Especially as Illustrated by the Data Obtained by Dr. G. Baur in the Galápagos Islands," *Proceedings of the Academy of Natural Sciences of Philadelphia* 48 (1896), 395–460. Information about the extensive loss of bulimulid species owing to introduced animals and plants may be found in Christine E. Parent and Robert P. Smith, "Galápagos Bulimulids: Status Report on a Devastated Fauna," *Tentacle* (January 2006), 24–26.

## Appendix 6

The adaptive functions of the human appendix and its role in immune responses to gastrointestinal illnesses are discussed in Michel Laurin, Mary Lou Everett, and William Parker, "The Cecal Appendix: One More Immune Component with a Function Disturbed by Post-industrial Culture," *The Anatomical Record* 294 (2011), 567–79.

# Acknowledgments

The origin of the Darwin Bear story goes back nearly thirty years ago, when I first wrote it as a birthday present for my godson, Michael Patrick Beatty, who loved Darwin Bear as much as a boy could possibly love a bear. Over the ensuing years, as I expanded the original manuscript and added new illustrations, many friends kindly advised me about the best way to present the Galápagos Bears' remarkable story. During this extended literary process, the book evolved from what had begun as a story to be told to a child into a didactic fantasy intended largely for adults who might wish to learn something about Darwin and evolutionary theory, including some of the ingenious approaches to scientific thinking that made Darwin as well as the Darwinian revolution so successful. I am particularly grateful for valuable advice and feedback from Mark Adams, Johannah Barry, Roslyn Cameron, James Costa, Frederick Crews, Jesa Damora, William Durham, Larry Gonick, Victoria Gray, Roald Hoffmann, Sarah Blaffer Hrdy, Heinke Jäger, Matthew James, Jill Key, Sonia Kleindorfer, Peter Kramer, Swen Lorenz, Katie Noonan, Jeanne O'Connell, Kari Olila, Lillian Park, Elizabeth Peele, Hans van Poelvoorde, Duncan Porter, Favio Rivera, Caleb Sherman, Dan Sherman, David Steadman, Anje Steinfurth, Carrie Toth, Alan Tye, Melissa Wells, and Mark Moffett (whose Galápagos friend, El Pingüino, keeps trying to insert himself into the story and has succeeded in doing so on page 39).

Malcolm Kottler's keen eye for detail regarding the history of evolutionary biology, Darwin scholarship, and important antiquarian distinctions about the various editions of Darwin's works, has provided an immeasurable contribution to the story. Sarah Lippincott and John Woram also provided greatly valuable editorial advice. I am especially indebted to Marissa Moss, an award-winning author and publisher of Creston Books, for her consummate editing of the manuscript and for her uncanny ability to think like a Galápagos bear.

The kind support of Marion B. Stroud and the Arcadia Summer Arts Program in 2003 allowed me to revise parts of this story in a most conducive environment characterized by inspiring company, spectacular landscapes, and a plentiful supply of wild blueberries on Mount Desert Island, Maine. Devin Shermer provided imaginative artistic suggestions for some of my drawings, and Pat Linse also gave me valuable artistic advice. I also wish to thank Fearn Cutler de Vicq not only for her skills as a book designer, but also for her assistance with the publication more generally.

I am enormously gratified by the enthusiastic support given for publication of this story by Wijnand Pon, Maas Jan Heineman, John Loudon, and the COmON Foundation in the Netherlands. Together with Mascha de Vries and Rubinstein Publications in the Netherlands, they first made it possible for the bears' story, together with their tireless efforts to save the Galápagos Islands from the scourge of invasive species, to be disseminated in English, Spanish, and Dutch. The COmON Foundation also deserves commendation for its generous financial support for the Galápagos Verde 2050 project, which has taken a lead in replanting native species on some of the islands where feral animals and human habitation have decimated much of the vegetation. In particular, I thank Patricia Jaramillo, head of the Galápagos Verde 2050 project, for her exemplary leadership and dedication to Galápagos conservation, as well as for her assistance with my own ecological research in these islands. Special thanks also to the Charles Darwin Research Station for allowing the bears to sleep in the rafters of the library whenever they are in Puerto Ayora.

The new American edition of this book, published by Blast Books, owes itself in significant part to Rosamond Purcell's many years of enthusiastic support for this work. In addition, I am grateful to Laura Lindgren and Ken Swezey at Blast Books for their keen interest in making the book more widely available, and for Laura's insightful editorial suggestions and refinements to the final book design for this new edition. A special thanks goes to my friend and fellow Darwin enthusiast Michael Shermer for providing a new foreword to the book, and for supporting, over the last three decades, my research efforts in many different domains.

My first visit to the Galápagos Islands as a twenty-one-year-old Harvard undergraduate changed my life. This visit was galvanized in part by having recently attended Edward O. Wilson's captivating course on evolutionary biology. Owing to his infectious love of the natural world and his explications of its endless evolutionary complexities, Ed was the single-best lecturer I encountered as an undergraduate. While I was taking this course, Ed provided his enthusiastic support for a documentary film expedition I was organizing to retrace Darwin's voyage on the *Beagle*, and he even offered funds from his course budget to help cover our expenses. Over the ensuing years, Ed has been an inspiration to us all for his brilliant writing on the subject of ants, biogeography, evolutionary theory, and conservation issues—topics that have earned him a host of prestigious scientific awards, along with two Pulitzer Prizes for General Nonfiction, as well as the enduring respect of his fellow scientists. *Darwin and His Bears* is dedicated to him with heartfelt thanks for the inspiration he has provided not only to key parts of the Darwin Bear story itself but also to my own academic career.

# Index

The author, on Isabela Island, using repeat photography in the Galápagos Islands to document ecological changes over the last three decades inside Volcán Alcedo's caldera (2008). Active fumaroles can be seen in the background. Photograph by Eric Rorer

# About the Author

FRANK J. SULLOWAY is an adjunct professor in the Department of Psychology at the University of California, Berkeley, and also a member of the Institute of Personality and Social Research at the same university. He has a Ph.D. in the history of science from Harvard University (1978) and is a MacArthur Fellow. His book *Freud, Biologist of the Mind: Beyond the Psychoanalytic Legend* (1979) provides a radical reanalysis of the origins and validity of psychoanalysis and received the Pfizer Award of the History of Science Society.

Dr. Sulloway has also published extensively on the life and theories of Charles Darwin. His various researches have taken him to the Galápagos Islands seventeen times, beginning with efforts to retrace Darwin's route there and to understand how these islands affected Darwin's thinking. Among other topics, he has conducted research on ecological disturbances in the Galápagos since Darwin's visit, and, together with Sonia Kleindorfer and other collaborators, he has published extensively on the evolution and behavioral ecology of Darwin's famous Galápagos finches.

For the last two decades, Dr. Sulloway has also employed evolutionary theory to understand how family dynamics affect personality development, including that of creative geniuses. He has a particular interest in the influence that birth order exerts on personality and behavior. In this connection, he is the author of *Born to Rebel: Birth Order, Family Dynamics, and Creative Lives* (1996). Dr. Sulloway's researches on birth order and family dynamics have been featured on a variety of national television shows, including *Nightline*, the *Today Show*, *Dateline NBC*, *Charlie Rose*, and *The Colbert Report*.

Dr. Sulloway has been the recipient of fellowships from the Institute for Advanced Study (Princeton, New Jersey), the Miller Institute for Basic Research in Science (University of California, Berkeley), the National Science Foundation, the John Simon Guggenheim Foundation, and the Center for Advanced Study in the Behavioral Sciences (Stanford, California). In addition, he is a Fellow of the American Association for the Advancement of Science, the Association for Psychological Science, and the Linnean Society of London. He is also a recipient of the Golden Plate Award of the American Academy of Achievement (1997), for which he was nominated by past recipients Francis Crick, Stephen Jay Gould, and Edward O. Wilson. He currently lives in Berkeley, California.